AF

PERGAMON INTERNATIONAL LIBRARY
of Science, Technology, Engineering and Social Studies
The 1000-volume original paperback library in aid of education,
industrial training and the enjoyment of leisure
Publisher: Robert Maxwell, M.C.

HYDROGEN POWER

An Introduction to Hydrogen Energy and its Applications

THE PERGAMON TEXTBOOK
INSPECTION COPY SERVICE

An inspection copy of any book published in the Pergamon International Library will gladly be sent to academic staff without obligation for their consideration for course adoption or recommendation. Copies may be retained for a period of 60 days from receipt and returned if not suitable. When a particular title is adopted or recommended for adoption for class use and the recommendation results in a sale of 12 or more copies, the inspection copy may be retained with our compliments. The Publishers will be pleased to receive suggestions for revised editions and new titles to be published in this important International Library.

HYDROGEN POWER

An Introduction to Hydrogen Energy
and its Applications

L. O. WILLIAMS

*Manager, Hydrogen Programs, The Aerospace Corporation, Germantown,
Maryland 20767, USA*

PERGAMON PRESS

OXFORD · NEW YORK · TORONTO · SYDNEY · PARIS · FRANKFURT

U.K.	Pergamon Press Ltd., Headington Hill Hall, Oxford OX3 0BW, England
U.S.A.	Pergamon Press Inc., Maxwell House, Fairview Park, Elmsford, New York 10523, U.S.A.
CANADA	Pergamon of Canada, Suite 104, 150 Consumers Road, Willowdale, Ontario M2J 1P9, Canada
AUSTRALIA	Pergamon Press (Aust.) Pty. Ltd., P.O. Box 544, Potts Point, N.S.W. 2011, Australia
FRANCE	Pergamon Press SARL, 24 rue des Ecoles, 75240 Paris, Cedex 05, France
FEDERAL REPUBLIC OF GERMANY	Pergamon Press GmbH, 6242 Kronberg-Taunus, Hammerweg 6, Federal Republic of Germany

First edition 1980

British Library Cataloguing in Publication Data

Williams, L. O.
Hydrogen power. - (Pergamon international library).
1. Hydrogen as fuel
I. Title
665'.81   TP359.H8   80-40434
ISBN 0 08 024783 0 (Hardcover)
ISBN 0 08 025422 5 (Flexicover)

In order to make this volume available as economically and as rapidly as possible the author's typescript has been reproduced in its original form. This method has its typographical limitations but it is hoped that they in no way distract the reader.

Printed and bound in Great Britain by
William Clowes (Beccles) Limited
Beccles and London

CONTENTS

ACKNOWLEDGEMENT

The author gratefully acknowledges the encouragement and helpful comments of the technical reviewer, T. Nejat Veziroglu, President of the International Association for Hydrogen Energy; the editorial efforts of Barbara Breslin of The Aerospace Corporation in making this book more consistent and readable; and the efforts of Diana Payne, Dolores Michlik, and Betty Viverette of the Aerospace Word Processing Center for producing the excellent quality final copy.

LIST OF FIGURES AND TABLES

Chapter 1

INTRODUCTION AND HISTORICAL BACKGROUND

Hydrogen is the most plentiful substance in the universe. It makes up about three-fourths of the matter of the universe, with helium making up most of the remainder, and the heavier elements so important to the earth and life amounting to only about 1 percent. The sun is still in the early stages of its life so it, too, is still about 75 percent hydrogen.

Deep in the interior of the sun, nuclear reactions are taking place that are converting hydrogen to helium and releasing the enormous energy necessary to keep the sun shining. The reaction is thought to be

$$4 \text{ hydrogen atoms} \longrightarrow 1 \text{ helium atom}$$

Hydrogen atoms have a mass of 1.67329×10^{-24} grams; helium atoms, a mass of 6.64553×10^{-24} grams. The mass of the four hydrogen atoms that react to produce a helium atom is therefore greater than the mass of the helium atom that is produced; the difference is converted to energy when the reaction takes place:

$$4(1.67329 \times 10^{-24}) - 6.64553 \times 10^{-24} = 4.761 \times 10^{-26} \text{ grams}$$

This difference, 0.711 percent of the original mass of the four hydrogen atoms, is the source of all the energy in the universe. The stars and galaxies shine because of this 0.711 percent mass that is lost when four hydrogen atoms are converted into a helium atom. This may not sound like a large amount of energy when expressed in this manner, but the appearance is deceiving. By application of Einstein's famous expression,

$$E = mc^2$$

1

where E is energy, m is the mass converted, and c is the velocity of light, the energy produced by the conversion of 1 gram of hydrogen into helium is

$$E = (1 \times 0.00711) \times (2.9979 \times 10^{10})^2 = 6.3900 \times 10^{18} \text{ ergs}$$
$$6.3900 \times 10^{18} \text{ ergs} \times 10^{-7} = 6.3900 \times 10^{11} \text{ joules}$$

To obtain the same amount of energy by burning fuels with oxygen, it would be necessary to consume 4,580,000 grams of hydrogen, 15,270,000 grams of fuel oil, and about 23,000,000 grams of coal. This primary hydrogen energy flowing through the universe not only drives all the astrophysical processes but, in the ultimate analysis, is the source of life and all the energies that we use here on earth.

Millions of years ago, hydrogen reacted to produce helium, and the energy released from this reaction diffused out to the surface layers of the sun, where the radiation energy escaped the sun's confines. Most of that energy was radiated out into the black depth of interstellar space, but a small pittance reached the earth, only about 5×10^{-10} fraction of the total solar energy. At the most, a few percent of this energy was captured by the plants growing on the earth. Most of what was captured was lost as heat when the plants died and decayed into the original substances of the air and soil, but a small percentage of that energy was trapped by the geological processes on the earth and converted into the fossil fuels which are now the basis of world industry.

Modest amounts of energy come from other sources, but all of these can ultimately be traced back to a point where the energy came from the fusion of four hydrogen atoms into an atom of helium. Hydroelectric power is the product of the evaporation of water into the atmosphere by the action of sunlight, followed by the condensation of the water to rain, some of which falls on land elevated above sea level. This is the youngest form of hydrogen power. The hydrogen atoms that combined to produce the energy to evaporate the water combined only a few thousand years ago, deep in the heart of the sun. Geothermal and nuclear fission are the oldest forms of hydrogen power. Geothermal energy is partly energy left over when the gas cloud from which the solar system was formed was compressed to form the sun and planets; part of it is derived from the slow decay of radioactive elements mixed with the substance of the body of the earth. The energy of radioactive elementsand the nuclear fission energy that can be released from them was derived from the fusion of hydrogen atoms in some star that was born, lived, and died before the formation of our solar system. When this ancient star had used most of the hydrogen in its core, it contracted and got hotter and hotter. As it neared its end, complex nuclear reactions generated the heavier elements, from iron to uranium. At the end of its life, a cataclysmic collapse caused much of the unreacted hydrogen to be converted to helium in an explosive reaction. This stupendous explosion, a supernova, blew the star apart and distributed the heavy elements back out into the void of space. These heavy elements were later incorporated into the cloud of gas from which the sun and planets formed, bringing the heavy elements synthesized in the death of a star to the formation of the earth. It can be seen from this that the atomic energy released from the fission of uranium is just a very old and altered form of the energy originally released by the conversion of four hydrogen atoms into one atom of helium.[1]

Even though hydrogen makes up 75 percent of the universe as a whole, it is a relatively rare element of the earth; it is estimated that the average concentration in the surface layers of the earth is only about 0.14 percent. The gravity of the earth is relatively low, and earth is sufficiently near the sun that the heat from the sun caused the escape of most of the hydrogen that was present when the earth was formed. The similar-sized, but hotter planet Venus seems to have even less hydrogen, and Mercury probably has almost none.

Hydrogen is the tenth most abundant element, lying between titanium and phosphorous on the abundance list.[2] Because it is a major constituent of the thin layer of water with which the planet earth is covered, it is quite accessible for use by mankind.

Research efforts in most of the industrialized nations of the world are directed at discovering how to react hydrogen to release energy by the formation of helium. These efforts are making progress, and it is thought that ultimately it will be possible to use these reactions to produce the energy needed to sustain modern civilization on a world-wide basis. The dream of abundant energy will come about relatively soon; just how soon depends on the vigor with which the research is performed.

The intent of this book, however, is not to discuss the development of hydrogen fusion for the production of energy, but to explain how hydrogen is currently produced, used, and handled and to show that the use of chemical hydrogen power has enormous advantages as an energy storage, transport, and use medium. In this application, hydrogen can be thought of as a kind of storable, portable electricity. It can be produced by diverse sources, ranging across the whole spectrum of primary energy sources from solar energy harvested as it is received, to secondary solar energy in the form of wind and water power, to fossil solar energy in the form of hydrocarbon fuels and nuclear fission energy. Hydrogen and electricity both have this capability of being produced from almost any source, but, unlike electricity, hydrogen can be stored in a material form for use at a later time, or perhaps as a fuel for vehicles. Over the next 20 to 40 years, the use of fossil fuels will greatly diminish because of environmental reasons as well as exhaustion of the supply,[3] and the probable candidate to take the place of these fuels for stored and transportation energy is hydrogen.

Hydrogen was undoubtedly produced by many of the early alchemists when they dissolved metals in acids, but it was not isolated until 1766 by Cavendish. He did not recognize it as a new substance but called it "flammable air." In 1783, Lavoisier identified hydrogen as a constituent of water and gave it the name hydrogen, meaning water-former. Little use was made of hydrogen until the middle of the 19th century, when coal gas came into use as a means of transmitting part of the energy content of coal through pipelines as a gas.

When coal is heated in a sealed vessel (pyrolized), a gas is given off that contains about 20 percent hydrogen and a number of light hydrocarbons. It was found that this gas could be used easily and relatively cleanly for cooking and lighting. After the coal had been pyrolized, the resultant solid, coke, was used as a special-purpose fuel in applications in which the volatile fraction of coal caused problems; notably, in the reduction of iron oxide ore to metallic iron in the blast furnace. During this period, the demand for gas grew at a faster rate than the demand for the by-product coke, so methods of improving the yield of gas in the coal processes were investigated. It was discovered that water injected into a bed of hot coke would produce a gas that was nearly an equal mixture of hydrogen and carbon monoxide, both combustible gases usable as fuel. This gas came to be called water gas.

Water gas burned cleanly, was reasonably easy to handle in the same equipment that had been used to handle coal pyrolysis gas, but had several shortcomings as compared to coal pyrolysis gas: The heat obtained by burning a volume of this gas was only about half that obtained from the combustion of the coal gas, and the flame was invisible and produced almost no light. In addition, the 40 to 50 percent carbon monoxide present in the gas made it extremely hazardous to use in the home. A gas light or stove burner that was inadvertently turned on, or left on, could release, in a matter of minutes, sufficient carbon monoxide to poison the inhabitants of the household. Carbon monoxide is dangerous to health at concentrations higher than 50 parts per million by volume and rapidly fatal at levels of 500 to 1000 parts per million.

Natural gas is mostly methane and, like hydrogen, it is not poisonous in the slightest and can only harm if its concentration is so high that it reduces the concentration of the oxygen in the air sufficiently that it can smother. The folklore that one can commit suicide by breathing the gas from a gas stove arises entirely from the past use of water gas, with its high carbon monoxide content, in home cooking stoves. It is very difficult to breath sufficient methane or hydrogen by holding your head over a gas stove to harm yourself. The leaking gas will become concentrated enough to explode, about 5 percent, long before it becomes concentrated enough to smother you, 40 to 70 percent.

Despite the real hazard of carbon monoxide, water gas was used for home cooking to some extent until 1950, when the expanding distribution of the safer and higher energy content natural gas displaced it from this market in the United States. Modest amount of hydrocarbons were added to water gas to make the flame luminous, and odorous substances added to aid in the detection of leaks before it was fed through the pipe to the home consumer. This mixture, containing 40 to 50 percent hydrogen, was used from about 1850 to 1950 as a home delivery method for convenient energy.

The use of pure hydrogen as an energy carrier was probably first mentioned in 1870 by Jules Verne in his story "The Mysterious Island." In this book, Verne writes "I believe that water will one day be employed as fuel, that hydrogen and oxygen which constitute it, used singly or together, will furnish an inexhaustible source of heat and light, of an intensity of which coal is not capable. I believe then that when the deposits of coal are exhausted, we shall heat and warm ourselves with water. Water will be the coal of the future." Verne recognized, even then, what we are having such a difficult time convincing people today: Fossil fuel resources are finite, and the fossil fuel age will come to an end; in fact, we are already beginning to see the end.

In 1923, Haldane presented a talk entitled "Daedalus or Science and the Future,"[4] in which he suggested that, at some time in the future, when fossil fuels were used up, England would obtain its power from the wind and convert it to hydrogen for storage and transportation. He included in his prediction that the hydrogen would be stored as a liquid in vacuum-jacketed storage vessels buried under the ground. This stored gas could be used to produce electrical energy when the wind did not blow strong enough to fulfill the need.

At the same time that Haldane was suggesting the use of hydrogen as a general-purpose fuel, there was a significant use of the pure gas in the lighter-than-air ships. During the First World War, both sides in the conflict used tethered ballons as observation platforms from which to view the activities behind the other's line. After the war, the commercialization of this concept proceeded at a good pace for several years, with a numberof ships being built, tested, and used in limited service. Until helium was obtained in quantity from natural gas in the early 1930s, hydrogen was the only gas used. Of the ships using hydrogen, the Graf zeppelin was the ultimate expression of this technology. It made its first transatlantic crossing in 1928, carried 13,110 passengers on that route during its 9-year service life, and was decommissioned in 1937 because of age.[5]

The Hindenburg, designed to use helium as the lifting gas, was completed in 1936. Because of the world situation at that time, the United States, with the only supply of helium, refused to sell it to the Hindenburg's German operators. The Hindenburg operators then made the fatal decision to use flammable hydrogen, even though the ship was not designed to operate with a flammable gas. A number of jury-rigged safety precautions were taken, and the ship was operated successfully until 1939.[5] In May 1939, it caught fire and burned on its arrival in the United States at Lake Hurst, New Jersey. It is conjectured that the cause of the fire was either leaking hydrogen, ignited by a static electricity discharge, or deliberate sabotage. As air disasters go, it was not a severe event. Of the 62 persons onboard, about half escaped without harm; many died as a result of attempting to jump to safety from too great a height.

The Hindenburg event was covered live by the radio, and there were a number of news photographers and newsreel cameramen present. As a result of this immediate and total coverage, almost everyone became familiar with the event, and this great familiarity caused the reputation of hydrogen as a very dangerous substance to grow out of proportion to reality. Hydrogen, like gasoline, natural gas, and other fuels, can be dangerous if used improperly; however, if handled with proper knowledge and care, it presents no greater hazard than other fuels, and possibly less.

The R101 airship, developed in England, used a unique technique to increase the range or the carrying capacity of the ship. As an airship travels, the hydrocarbon fuels are consumed and the ship becomes lighter by the amount of fuel consumed. To maintain proper neutral buoyancy, it is necessary to vent hydrogen gas to reduce the lifting capacity of the ship by an amount equal to the weight of the consumed fuel. If the hydrogen to be vented were burned in the engines, the amount of fuel required would be reduced by an amount equivalent to the heat of combustion of the hydrogen consumed. This scheme of burning the hydrogen slated to be vented would reduce the necesary fuel load to accomplish a given mission by about 20 percent. This scheme was adopted and used on some of the flights of the R101.[6] During this same time period, Mr. Rudolph Erren, who worked on the German zeppelin project, was experimenting with the production and combustion of hydrogen in internal combustion engines for the same application.[7,8]

The renowned I.I. Sikorski, developer of the first helicopter, reviewed the advantages of the use of hydrogen as aircraft fuel in 1938.[9] He suggested that "If a method of safe and economical production and handling of liquid hydrogen were developed for use as a fuel, this would result in a great change, particularly with respect to long range aircraft. This would make possible the circumnavigation of the earth along the equator in a nonstop flight without refueling. It would also enable an increase in the performance of nearly every type of aircraft."

From 1948 through 1955, R.O. King and his colaborators at the University of Toronto conducted studies on the use of hydrogen in ordinary internal combustion engines.[10,11] King found that hydrogen could be used in ordinary engines if the compression ratio was maintained under about 7 and if the engine was kept clear of deposits that could trigger preignition. His studies were the first modern scientific investigation of the use of hydrogen as a potential fuel for the replacement for gasoline.

In the 1950s, there was interest in the construction of a special-purpose airplane that had an extremely long range and could fly at very high altitude. This aircraft was to be used as a photoreconnaissance plane by the U.S. Air Force. The engineers at the NAC Lewis Laboratory (now NASA Lewis Laboratory) performed the systems analysis necessary to provide the design concept for this aircraft and found that hydrogen was the only fuel that would let them attain the desired altitude and range.[12] Pratt and Whitney Corporation then undertook a development program leading to the modification of an existing turbojet engine to burn hydrogen.[13] The program was a complete technical success, and no significant problems were encountered. Ultimately, a Martin B57 bomber was flown with one engine coverted to the combustion of hydrogen.[14] Although the technical feasibility of the conversion of aircraft jet engines to hydrogen was thus demonstrated, the program was discontinued because other methods were developed for obtaining the desired photographs.

The Institute of Gas Technology has long been a proponent of the use of hydrogen as a general-purpose fuel. Its interest stems from the clear knowledge that eventually the supplies of natural gas will be depleted and its supporters, a consortium of North American gas producers and pipeline owners, will have a highly developed pipeline network covering North America with no gas to transport. A shift to the transport of hydrogen would seem a natural and desirable use for this enormous capital investment. In 1968, the Institute exhibited a "Home For Tomorrow," which contained working models of hydrogen-fueled catalytic appliances such as stoves (Figure 1-1), grills, and space heaters.[15]

Figure 1-1.   A Catalytic Hydrogen Stove (Courtesy of the
Institute of Gas Technology, Chicago,
Illinois).

During the 1960s, the hydrogen-oxygen fuel cell was developed. The fuel cell is a form of battery, differing in that the chemicals that react to produce the electric current can be continuously supplied and the product of the reaction can be continuously removed. In ordinary batteries, one plate of the battery, usually a metal, gives up electrons to the external circuit and, in doing so, is changed to a solid salt by reaction with the electrolyte. The other plate accepts the electrons and is converted into a second solid. When the plates of the battery are used up, the battery is dead and cannot produce any current. In the fuel cell, the hydrogen gives up the electrons and the oxygen accepts them to produce hydrogen ions and hydroxyl ions, respectively. These react in the electrolyte to produce water, and the water can be removed from the cell as a liquid or a gas. Because the products are continuously removed and the reactants can be continuously supplied, the fuel cell is a battery that will never wear out as long as it is supplied with hydrogen and oxygen.

The fuel cell was developed by the National Aeronautics and Space Administration into a highly reliable electrical generating device, albeit expensive, and was used to provide the electric energy used on the Apollo missions to the moon. In this application, the cells not only supplied the power needed but, as a bonus, supplied the drinking water as well. The Apollo fuel cells used platinum catalysts and were therefore relatively expensive to manufacture, but much development has taken place in the years since the Apollo missions and future fuel cells of lower cost will likely play a large part in the use of hydrogen as a fuel.

Around 1970, a number of systems engineers looking for solutions to the impending fuel crisis and pollution caused by the wide-scale use of fossil fuels independently reasoned that an economy based on the use of hydrogen as a universal energy carrier could offer solutions to both of these problems. The author,[16] Dr. Derek P. Gregory of the Institute of Gas Technology,[17] William J.D. Escher[18] (an independent consultant), Cesare Marchetti of the Euratom Joint Research Center[19], Lawrence Jones[20] of the University of Michigan, and J. O'M. Bockris of Flinders University, Australia,[21] all published system scenarios describing how a significant portion of the economy of an industrialized society could be converted to the use of hydrogen and the benefits that would acrue from such a conversion. These scenarios, of course, all had a rather futuristic flavor and disagreed on a number of details about how to implement the adoption of hydrogen as a fuel and when it might be possible to convert a significant portion of the energy system to hydrogen. They all agreed that there would be enormous benefit to society and to the environment of the whole earth if the conversion were made.

During the early 1970s, a number of articles appeared in magazines such as The Saturday Evening Post,[22] Business Week,[23] Time,[24] The American Legion Magazine,[25] Fortune,[26] Science News,[27] and Scientific American[28] repeating the theme that it would be highly beneficial if industrial societies would convert to the use of hydrogen for a general-purpose fuel. Each author had a different notion of the mix of energy sources that could be used as the primary energy source for the production of the hydrogen, but in general they agreed on the features of the energy system that should be adapted and the advantages that could be gained.

John O'M. Bockris first referred to the concept of the use of hydrogen as a general-purpose fuel as the "hydrogen economy." This designation has become quite popular but has caused some confusion in that it is often equated with the thought that hydrogen is somehow going to replace money as a medium of exchange. The author prefers the designation "hydrogen power," used in the same sense that "electric power" is used to describe the use of electricity as an energy carrier.

Many of the hydrogen proponents encountered a stereotyped reaction when they spoke of using hydrogen power. Inevitably, someone in the audience would comment on hydrogen being very dangerous, using the Hindenburg as an example. The comment was not very relevant, but it had the effect of stirring up a big debate over what happened to the Hindenburg. In May 1972, a group of hydrogen advocates formed the H_2indenburg Society, a very informal group dedicated to the safe use of hydrogen as a fuel. The group elected Dr. Derek Gregory of the Institute of Gas Technology president and William Escher of Escher Technology Associates secretary. There were no scheduled meetings, no minutes kept, no dues, no publications, and no way to officially join. The purpose was to promote the use of hydrogen, keep the members in informal contact, and attempt to convince people that what happpened to the Hindenburg was not a direct reflection on the safety or the potential for hydrogen as a fuel.

In April 1973, Dr. Simpson Linke of Cornell University conducted an International Symposium and Workshop on the Hydrogen Economy. The symposium was sponsored by the National Science Foundation's Research Applied to National Needs (RANN) Program. Dr. Linke hoped to get together, for 1 day, 30 or so of the researchers and engineers who had been working to promote interest in hydrogen as a fuel, to discuss progress and plans. He found that there was far more interest in the subject than he had anticipated, resulting in a 3-day symposium with 134 participants. Discussion showed that, although there was a great deal of interest in the hydrogen power concept, the lack of near-term payoff for the research and development made it very difficult to obtain funds. This problem still plagues the efforts to promote hydrogen power. There seems to be near unanimous agreement among scientists and engineers that someday hydrogen power will be a major contributor to the function of civilization, but because the concept cannot supply immediate answers to today's problems, it receives only modest financial support.

In March 1974, a symposium[29] was organized by T. Nejat Veziroglu, Professor and Chairman, Department of Mechanical and Industrial Engineering, University of Miami, and William Escher. This conference, "The Hydrogen Economy Miami Energy Conference," attracted more than 700 participants from over 30 nations. At the conference, it became very clear that there was worldwide and sustained interest in the concept of hydrogen power. Dr. Gregory and Mr. Escher were inundated with requests to join the H_2indenburg Society, requests they were unable to handle because no formal society existed. On the third day of the symposium, Dr. Gregory, Dr. Veziroglu, Mr. Escher, and the author agreed to see if there was sufficient interest among the participants to found a more formal organization for the promotion of hydrogen power. That evening, 75 conference participants met and, after much discussion about the H_2indenburg Society, its place, and the intent for the future, decided that the Society should not be continued and that a new society should be formed. There were the problems of finding someone who had the time necessary to form an organization, writing a charter, establishing bylaws, and preparing membership material, just as a start. Few of the people present had the capability to devote the time necessary for the formation of the new society. After some debate, Dr. Veziroglu agreed that, if the rest of the group would support him by helping with the formation, paying dues, and recruiting new members, he would be willing to take on the task of forming the new organization.

Over the next several months, Dr. Veziroglu succeeded in forming this new organization and, with the help of the founding members, launched the International Association for Hydrogen Energy*. This organization has sponsored a world conference on hydrogen every other year since it was founded: The 1976 conference was held in Miami, Florida; the 1978 in Zurich, Switzerland, and the 1980, in Tokyo, Japan. It publishes a bimonthly journal and serves as the focal point for hydrogen power research and development communications throughout the world. The association has witnessed a significant growth in its membership and in the funded research and development that has taken place throughout the world during its existence, but the elusive goal of developing a substantial demonstration of the use of hydrogen power for a substantive portion of the economy has not been achieved and remains a major goal for the future.

---

* P.O. Box 248266, Coral Gables, Florida 33124, U.S.A.

Chapter 1 References

1. David Clarke, Superstars, J.M. Dent, 1979.

2. Handbook of Physics and Chemistry 55 Edition, The Chemical Rubber Co., Cleveland, Ohio.

3. Owen Phillips, The Last Chance Energy Book, The Johns Hopkins University Press, 1979.

4. J.B.S. Haldane, Daedalus or Science and the Future, Kegan, Paul, Trench, Trubner and Company Ltd., London, 1924.

5. Jean, All the Worlds Aircraft.

6. Proceedings of the Institution of Mechanical Engineers 141, pp. 386-424, London, 1939.

7. R.A. Erren, French Patent 804,226, October 29, 1936.

8. R.A. Erren and W.H. Campbell, Hydrogen From Off-Peak Power--A Possible Commercial Fuel," Chem. Trade J. 92; pp. 238-239, 1933.

9. I.I. Sikorski, "Science and the Future of Aviation," Steinmetz Memorial Lectures, Schenectady Section, A.I.E.E., March 1938.

10. R.O. King, W.A. Wallace, and B. Mahapatra, "The Nuclear Theory of Ignition," Canadian J. Research, 26F: pp. 264-276, University of Toronto, 1948.

11. R.O. King and M. Rand, "Oxidation Decompositon Ignition and Detonation of Fuel Vapors and Gases. XXVII. Hydrogen Engine," Canadian Journal of Technology 33, pp. 445-469, 1955.

12. Lewis Laboratory Staff, "Hydrogen for Turbojet and Ramjet Powered Flight," NACA, RM E57D23, April 26, 1957.

13. R.C. Mulready, Liquid Hydrogen Engines, Chapter 5, Technology and the Uses of Liquid Hydrogen, Pergamon Press, McMillan Company, New York and London, 1964.

14. S. Weiss, "Hydrogen Fueled Aircraft," <u>Cryogenics and Industrial Gases</u>, pp. 13-19, November/December 1974.

15. J.C. Sharer and J.B. Pangborn, "Utilization of Hydrogen as an Appliance Fuel Hydrogen Energy," pp. 875-887 (Proceedings of the Hydrogen Economy Miami Energy Conference, Plenum Press, T.N. Verziroglu, editor, 1975).

16. L.O. Williams, "Plan for the Elimination of Pollution," <u>Design News</u>, January 5, 1970.

17. D.P. Gregory, "Hydrogen Economy," <u>Electrochemistry of Cleaner Environments</u>, Chapter 8, Plenum Press, J.O'M. Bockris, editor, 1972.

18. W.J.D. Escher, "Macro System for the Production of Storable, Transportable Energy from the Sun and the Sea," American Chemical Society, Div. Fuels Chem., Preprint 16, No. 4, pp. 28-47, April 10, 1972.

19. G. deBeni and C. Marchetti, "Hydrogen-Key to the Energy Market," Euro-Spectra 2, pp. 46-50, January 1970.

20. L.W. Jones, "Liquid Hydrogen as a Fuel for the Future," <u>Science</u> 174, pp. 367-370, October 22, 1971.

21. J.O'M. Bockris, "Hydrogen Economy," <u>Science</u> 176, p. 1323, June 23, 1972.

22. L.W. Jones, "Hydrogen, A Fuel to Run Our Engines in Clean Air," <u>Saturday Evening Post</u>, 1972, spring.

23. Staff, "When Hydrogen Becomes the World's Chief Fuel," <u>Business Week</u> 2247, p. 98, September 12, 1972.

24. Staff, "Fuel of the Future," <u>Time</u> 46, September 11, 1972.

25. H. Ardman, "A Novel Proposal for the Fuel of the Future," <u>American Legion Magazine</u>, July 1973.

26. L. Lessing, "The Coming Hydrogen Economy," <u>Fortune</u> 138, November 1972.

27. R.J. Trotter, "Is Hydrogen the Fuel of the Future?" <u>Science News</u> 102 (3), p. 46, July 13, 1972.

28. D.P. Gregory, "The Hydrogen Economy," <u>Scientific American</u> 228 (2), p. 13, January 1973.

29. T.N. Veziroglu, "<u>The Hydrogen Economy Miami Energy (THEME) Conference Proceedings</u>" pp. 925, University of Miami, March 1974.

Chapter 2

CHEMICAL AND PHYSICAL PROPERTIES OF HYDROGEN

Hydrogen is unique among the elements. It is characterized by a single electron valence electron, a property it shares with the alkali metals lithium, sodium, potassium, rubidium, and cesium. Unlike these metals, however, hydrogen shows little tendency to lose this electron in chemical reactions, preferring to pair it with another electron and form molecules such as H_2, HCl, and NH_3. The hydrogen atom lacks one electron of a completed shell, similar to helium, a property it shares with the halogens fluorine, chlorine, bromine, and iodine. But unlike the halogens, its electronegativity is so small that it can gain an electron only from the extremely electro positive metals. In many respects, hydrogen resembles the metallic elements, particularly in comparing the behavior of the solvated cation H^+ with the behavior of other solvated cations. Yet, although the cation appears metallic in its behavior, the majority of its properties and those of its compounds are more nearly like the nonmetallic elements. These apparent anomalies in behavior are reconcilable in terms of the unusual structure of the element, a single electron combined with a single proton as a nucleus with no intervening electron shells. These unique properties contribute to the chemistry of hydrogen that make it of such great interest as an energy carrier.

There are three isotopes of hydrogen, with atomic masses of 1.0079, 2.0142, and 3.0144; they are hydrogen (protium), deuterium, and tritium, respectively. In keeping with its unique characteristics, hydrogen is the only element that has separate names for its isotopes. Only hydrogen and deuterium are found in measurable quantities in nature; hydrogen abundance is 99.985 percent and deuterium, 0.015 percent. Most of the properties reported for natural hydrogen are those of protium. Deuterium, as a consequence of its low concentration, contributes negligibly to the gross properties of natural hydrogen.

15

Tritium is an unstable radioactive element with a half-life of 12.26 years. It decays, by the emission of a negative electron, to a very rare isotope, helium 3. By highly sensitive radioactive detection techniques, very small amounts of tritium can be detected in freshly precipitated water, the concentration being on the order of 1 atom of tritium for 10^{18} atoms of ordinary hydrogen. Natural tritium is produced by the action of cosmic rays on the atoms in the atmosphere. Tritium is quite important in the investigations leading to the release of fusion energy, but it is so rare that its properties and existence can be disregarded when considering the properties of hydrogen as an energy carrier.

The ratio of 2 between the masses of deuterium and hydrogen is much greater than the mass ratio between any other two isotopes, and the chemical behavior of deuterium is measurably different than that of hydrogen as a result. Deuterium was first isolated by the exhaustive electrolysis of large quantities of water. It was found that, when water is electrolyzed under suitable conditions, the lighter isotope, protium, is liberated roughly six times more readily than is deuterium[1]. Deuterium enters into all the reactions characteristic of ordinary hydrogen and forms completely equivalent compounds. However, the large mass difference between the two isotopes renders rates of equivalent reactions considerably different. In general, deuterium reacts more slowly and less completely than its lighter analog, and these properties characterize the behaviors of the corresponding compounds as well. In catalytic hydrogenations, deuterium reacts more slowly, but the nature of the resulting compounds is the same as with hydrogen. Deuterium is more slowly absorbed by metals that form hydrides by direct reaction. Aluminum carbide reacts with ordinary water to form methane with great rapidity, whereas deuterium oxide reacts quite slowly[2]. Ammonium salts dissolved in deuterium oxide (heavy water) react so as to replace all the protium in the ammonium ion with deuterium:

$$NH_4^+ + D_2O \longrightarrow ND_4^+ + H_2O$$

Some of the major properties of hydrogen, deuterium, and their oxides are listed in Table 2-1.

The chemistry and properties of deuterium are quite interesting but, due to its rarity, are of no direct importance in the consideration of hydrogen as an energy carrier. Deuterium is of great importance, however, as heavy water in the operation of the CANDU nuclear reactors built by Canada, in which heavy water serves as both the neutron moderator and a heat transfer medium.[3] In the future, it may become

Table 2-1. Properties of Hydrogen and Deuterium and Their Oxides

	H_2 (°K)	D_2 (°K)	H_2O (°C)	D_2O (°C)
Freezing Point	13.9	18.7	0	3.82
Boiling Point	20.38	23.59	100	101.42
Latent Heat of Fusion				
(calories/mole)	28.0	52.3	1435	1522
Latent Heat of Vaporization				
at 13.92°K (solid)	245.7	340.8		
at 100°K			9719	9960

important as one of the fuels of the hydrogen fusion, nuclear power-generating process, but it will not be considered further in the current discussion.

In 1927, Heisenberg[4] predicted from quantum mechanical considerations that two forms of hydrogen should exist, one in which the nuclear spins were parallel (ortho) and one in which they were opposed (para). A temperature-dependent equilibrium exists between the ortho and para forms, with the para form (opposed spins) being favored at low temperatures. Table 2-2 lists the concentration of the para form at equilibrium at a series of temperatures. The ortho form is at a slightly higher energy level than the para, and the conversion of the ortho form to the para releases a small amount of heat. The equilibrium is reached quite slowly, unless the conversion is aided with a catalyst.

When liquid hydrogen was first prepared, it was found to have a much higher boil-off loss rate than was predicted from the heat leak measurements made on the containers. This enhanced boil-off was caused by the slow heat release from the conversion of the ortho form to the para form. In all current liquid hydrogen production operations, the conversion of the ortho form to the para form is catalyzed, usually with magnetic iron oxide (Fe_3O_4), as the liquefaction is performed, so that the

Table 2-2. Equilibrium Concentration of Para Hydrogen

Temperature ($^{\circ}$K)	Percentage Para H_2
0	100.0
25	99.00
50	76.80
100	38.46
200	25.95
250	25.26
298	25.07
00	25.00

resultant liquid hydrogen is in the para form and shows the boil-off rate expected from the heat leak calculations performed on the containers.[5]

The hydrogen molecule can be dissociated by very high temperatures, such as those found in an electric arc operated at high current density, in the following manner:

$$H_2 \longrightarrow H + H$$

This reaction is extremely endothermic, requiring 104.18 kilocalories per mole of hydrogen (water decomposition requires 68.32 kilocalories per mole of water), and the resultant atomic hydrogen is a powerful reducing agent and reacts with great rapidity. Atomic hydrogen can also be produced by radiating hydrogen with ultraviolet radiation of short enough wavelength that the photons can supply the necessary energy to disrupt the molecule.

Atomic hydrogen reversion is catalyzed by contact with metals. When a stream of atomic hydrogen produced in an arc is allowed to impinge on a metal, the reversion takes place preferentially on the surface of the metal. The heat released at the surface of the metal is sufficient to melt almost all metals. This property of hydrogen is utilized in the atomic hydrogen torch used in welding very refractory metals.

Atomic hydrogen is an active reducing agent. It reduces the oxides and chlorides of copper, lead, bismuth, silver, and mercury directly to the free metal. Sulfur is converted to hydrogen sulfide. Alkali metal salts such as nitrates, nitrites, azides, cyanides, and thiocyanates are reduced to the free metal. Further exposure to the atomic hydrogen converts the alkali metals to the corresponding hydrides.[6]

Chemically combined hydrogen may be classified conveniently as covalent or anionic. The ionization potential of hydrogen is so large (13.595 electron volts = 313.6 kilocalories per gram mole) that truly cationic hydrogen in chemical combination is impossible. Hydrogen does combine with more electronegative elements (chlorine, iodine, etc.) and is thereby assigned a positive oxidation number, but such compounds are polar rather than ionic. Very pure water and very pure and dry liquid hydrogen chloride have very low conductances, indicating low levels of ionization. If the bonds in such compounds are sufficiently polar, they may be ruptured on contact with highly polar solvents to produce solvated protons and the anion characteristic of the original compound. Hydrogen chloride thus dissolves in highly polar water to form a solvated hydrogen cation:

$$HCl + Excess\ H_2O \longrightarrow H_3O^+ + Cl^-\ solution$$

Free protons can no more exist in these solvents than in pure compounds, thus purely cationic hydrogen is limited in its existence to discharge tubes and nuclear reactions.

Covalently bonded hydrogen is characteristic of the volatile hydrides such as methane (CH_4), silane (SiH_4), arsene (AsH_3), and the like. Anionic hydrogen is not unexpected. Upon gaining an electron, the hydrogen atom gains the two-electron, filled shell structure of helium, which is quite stable. The sizable electronegative difference between hydrogen and some of the metals reinforces the expectation of anionic hydrogen. These differences, however, are only large enough to permit electron transfer to hydrogen with very highly electropositive elements such as the alkali and alkaline earth metals. Binary compounds between hydrogen and such metals contain the negative hydride (H^-) ion. Because covalence greater than one is not possible for hydrogen (the stable shell contains only two electrons), coordination to bonded hydrogen does not occur.

By direct or indirect means, binary combinations of hydrogen with almost all the other elements can be prepared and are commonly referred to as hydrides. This, however, is not in strict keeping with common practice in nomenclature, for combinations with elements that are more electronegative than hydrogen are probably even more common than those with elements that are less electronegative. Because of the variety of elements appearing in combination with hydrogen, the hydrides are characterized by a wide variety of properties and are thus representative of several types of bonding. Early interest in hydrogen compounds of the nonmetals, particularly oxygen, the halogens, and carbon, lead researchers to attempt to synthesize and characterize a number of the other nonmetallic hydrides. The efforts of Stock[7] and his coworkers lead to the elucidation of the elemental composition of the hydrides formed by boron and silicon, and this in turn lead to further interest and work in elucidating the structures of other hydrides. Paneth[8] assembled the data on the diverse hydrides and prepared a classification system for the hydrides. His original classification indicated the salt-like hydrides (sodium hydride), the covalent hydrides (methane), and the metallic hydrides (palladium hydride). Although this classification provides insight into the various types of hydrides, the situation can be more accurately viewed as a continuous gradient of properties from the salt-like hydrides of the alkali metals to the purely covalent hydrides of carbon, with the hydrides of the other elements lying somewhere between. The metallic hydrides are treated separately as alloys of hydrogen and the metal, lending substance to the classification of hydrogen as a special and peculiar member of the alkali metal family.

The alkali and the alkaline earth metals form saline-type hydrides: LiH, NaH, KH, RbH, and CsH for the alkalies and BeH_2, MgH_2, CaH_2, and BaH_2 for the alkalines. The hydrides of the most electropositive lanthanides and actinides demonstrate the saline crystaline structure and high heats of formation characteristic of the saline hydrides (Table 2-3), but in some cases it is difficult to prepare these compounds with a stoichiometric composition. They most closely resemble the saline hydrides and can be classified with them without serious compromise. All the saline hydrides show a crystaline structure and can be prepared by the direct reaction of the pure metals with hydrogen at temperatures ranging from 150° to 600°C. Calcium will react with hydrogen at temperatures as low as 150°C, whereas lithium must be heated to temperatures of 600°C to cause a reaction with hydrogen. In preparing these hydrides, it is necessary to take great pains to ensure that the reaction is complete, for the initial hydride formed can protect the metal from further reaction with the hydrogen. Strong agitation, grinding, dispersion of the metal in an inert solvent, and high hydrogen pressures all favor the formation of hydrides in good yields.

Alkali metal hydrides possess the cubic crystal structure of the type exhibited by sodium chloride, giving further support to the anionic character of the hydrogen in these types of compounds. The alkaline metal hydrides show somewhat more complex structures, but it has been established that, in both alkaline and alkali hydrides, the lattice is ionic and contains negative hydride (H-) ions formed by the transfer of

Table 2-3. Heats of Formation of Some
Representative Hydrides

		Kilocalories mole
Saline	LiH	-21.61^a
Saline	NaH	-13.7^a
Saline	KH	-11.3^b
Polymeric	BeH_2	$+10.-^b$
Saline	MgH_2	$-18.-^b$
Saline	CaH_2	-45.1^a
Saline	SrH_2	-42.3^a
Interstitial	$LaH_{2.76}$	-40.09^a
Interstitial	$CeH_{2.69}$	-42.26^a
Saline	UH_3	-30.4^a

[a]Handbook of Physics and Chemistry, 59 Edition, The
Chemical Rubber Co., 1978.

[b]Siegel, B. and Schieler, L., *Energetics of Propellant
Chemistry*; John Wiley & Sons, Inc., New York, 1964.

electrons from the metal atoms to the hydrogen. For example, when lithium hydride is heated to fusion temperatures, it conducts an electric current and undergoes electrolysis. When this electrolysis is performed, the lithium is deposited at the cathode and hydrogen at the anode, in quantities consistent with electrochemical considerations.

The saline hydrides are not soluble in common solvents. With water and acids they react with the release of hydrogen gas to form the respective salt of the metal:

$$LiH + H_2O \longrightarrow H_2 + LiOH$$
$$CaH_2 + 2HCl \longrightarrow 2H_2 + CaCl_2$$

They do dissolve without reaction in molten halides, and sodium hydride is soluble in molten sodium hydroxide. Only the most stable of the hydrides, lithium hydride, can be melted without decomposition; it melts at 680°C. The alkali hydrides other

than lithium hydride undergo significant decomposition in the temperature range of 400°
to 500°C.   Lithium hydride and the alkaline earth metal compounds decompose at more
elevated temperatures.   The temperature of decomposition is influenced by the presence
of transition metal impurities; iron and nickel in particular lower the decomposition
temperatures.

At elevated temperatures, all the saline hydrides are powerful reducing agents.
Refractory metal oxides are reduced to the free metals and sulfates to the corresponding sulfides when they are pyrolized in mixtures with the hydrides.   Many organic substances undergo reductions when treated with the hydrides at modestly elevated temperatures.

Lithium, calcium, and strontium hydride are stable in dry air, but others may
ignite spontaneously, although sodium hydride is stable to ignition at temperatures as
high as 230°C.   The presence of traces of the unreacted metals may lower the temperature at which the hydride will react with air and, as it is difficult to drive the
hydriding reaction to completion, unreacted metal is often present.

In the past, hydrides have not been used in industry to any large extent.   Recently, their use has grown because of the development of reduction processes using
hydrides and a number of their complex derivatives to perform highly specific organic
chemical reductions.   For further details on these reactions, see the excellent mono-
graph by H.C. Brown.[9]

Covalent hydrides are known for all of the elements that are clearly nonmetals,
except the rare gases.   Aluminum, gallium, tin, lead, antimony, and bismuth form
hydrides that appear more nearly covalent despite the nominal classification of the
elements as metallic.   Except for boron, aluminum, and gallium, these elements give
compounds that have the composition expected on the basis of simple valence rules.
For boron and gallium, the simplest hydrides are the dimeric B_2H_6 and Ga_2H_6; for
aluminum, it is the polymeric $(AlH_3)_x$.   Diborane (B_2H_6) boils at -92.5°C, digallane
(Ga_2H_6) boils at 139°C with decomposition, and polymeric aluminum hydride (AlH_3)
shows no boiling point but begins decomposition at temperatures somewhat over 100°C.
These elements--with the exception of tin, lead, antimony, and bismuth--and the halogens form polynuclear hydrides in which two or more atoms of the nonmetal are directly attached to each other.   This property is most spectacular in the case of carbon, for which there are hundreds of thousands of this kind of compound known and an
enormously larger number that seem possible.   In silicon, this tendency is greatly reduced, and chains of silicon atoms greater than four or five seem impossible to prepare.   Boron forms a group of compounds that contain both boron-boron linkages and a

unique form of hydrogen bridge bond.   One of the most stable boron hydrides, decaborane, $B_{10}H_{14}$, has a structure in which the boron atoms are arranged in the form of a basket. Nitrogen forms only one compound in this category that can be isolated as a pure nitrogen hydride, hydrazine N_2H_4, a liquid boiling at 111^oC with properties similar to water.   Oxygen forms only hydrogen peroxide (H_2O_2), with an oxygen-to-oxygen bond. With the heavier nonmetals, this tendency is even more depressed, with P_2H_6 and H_2S_x being quite unstable.

A number of synthetic routes are used in the preparation of the covalent hydrides. Direct combination of hydrogen with the nonmetal is suitable for the hydrides of oxygen, nitrogen, and the halogens.   These reactions are all quite exothermic and, with the exception of the reactions of nitrogen and iodine, proceed with explosively rapid reaction rates.   Nitrogen and iodine react only at elevated temperatures.   Hydrolysis of an active metal (often magnesium) compound of the nonmetal is useful in the preparation of the hydrides of silicon, boron, carbon, nitrogen, phosphorous, sulfur, and such.   With boron and silicon, the reaction products are often complex mixtures. Calcium and aluminum carbides, calcium, aluminum, and magnesium nitrides react vigorously with water to yield the nonmetal hydrides.   Other examples of these reactions require dilute acids for the hydrolysis.   Calcium carbide hydrolyzes to produce acetylene (C_2H_2), and aluminum carbide produces only methane (CH_4). For the production of the hydrides of silicon and boron, the reactions of magnesium silicide or boride are the only methods of importance. Electrochemical reductions of certain solutions can produce hydrides; for example, the hydride of tin is obtained when sulfuric acid solutions of tin (II) sulfate are electrolyzed.   A method that has general applicability to the production of many of the nonmetallic hydrides is the treatment of ether solutions of the chloride of the nonmental with lithium aluminum hydride ($LiAlH_4$), an extremely powerful reducing and hydriding agent.

The hydrides of aluminum and boron are both able to coordinate a fourth hydrogen atom, a property that leads to a group of very active and yet relatively stable compounds.   Lithium hydride reacts with both boron and aluminum hydride to form lithium borohydride ($LiBH_4$) and lithium aluminum hydride ($LiAlH_4$), respectively.   The sodium and potassium analogues of these compounds can also be prepared.   Aluminum hydride and diborane can be combined to produce aluminum borohydride ($Al(BH_4)_3$), a liquid at room temperature.

Aluminum borohydride, silane, and some of the boranes have ignition temperatures that are below room temperature, making handling quite difficult.   Silane is the most extreme in this respect and will ignite spontaneously in air at temperatures as low as -100^oC.

The compounds of hydrogen and carbon, the hydrocarbons, are without question the most important group of chemical compounds known. First, and most important, they make up the chemical structures that make life possible. No other group of elements can form the extreme diversity of compounds that can be formed from hydrogen and carbon in conjunction with smaller amounts of oxygen, nitrogen, and many trace elements. This diversity of compounds is probably necessary to perform the many different functions required for life to exist. Although it cannot be said with absolute certainty, it is extremely probable that, if life is ever discovered elsewhere in the universe, it too will be found to be based on the chemistry of carbon and hydrogen. It is also likely that the hydrogen and oxygen compound, water, will be the solvent in which the reactions of the life-sustaining hydrocarbons will take place. The only other remote possibility for the solvent system would be ammonia, also a hydrogen compound. The chemistry of life systems and their associated hydrocarbon compounds is so rich that there is a special subdiscipline of chemistry, biochemistry, to encompass it. This discipline is an offshoot of what was originally the study of the compounds of life, organic chemistry. Today, organic chemistry tends to be more involved in hydrogen and carbon chemistry that is not involved in life.

The many hydrogen and carbon compounds of non-life organic chemistry are also very important to the maintenance of civilization. Virtually all fuels used fall into the category of non-life organic compounds. It is thought that these compounds originated from biological materials that were formed ages ago, buried, and altered by the internal heat and pressure under the ground. There are vestiges of their biological origins in the structure of these compounds today, but the majority of the compounds are organic structures not found in living organisms.

The polymeric substances that make up all the plastics, synthetic fibers, paints, and adhesives are synthesized from organic chemicals extracted from the fossil fuels. This area of hydrogen and carbon chemistry has grown since the 1930s into a major tonnage industry that has profound influence on the success of modern society. Use of these compounds as fuels is very irresponsible in that it will be far more difficult for future generations to replace them by synthesizing the chemicals than it would be for the current civilization to find new materials for fuels.

Because of the ability of carbon, combined with hydrogen, to link with itself at four distinct valence attachment points, the diversity for possible compounds is magnificent. Tables are available that list several hundred thousand different compounds for which some of the physical and chemical properties have been determined and verified. The recent literature in organic chemistry lists millions of compounds for which a few

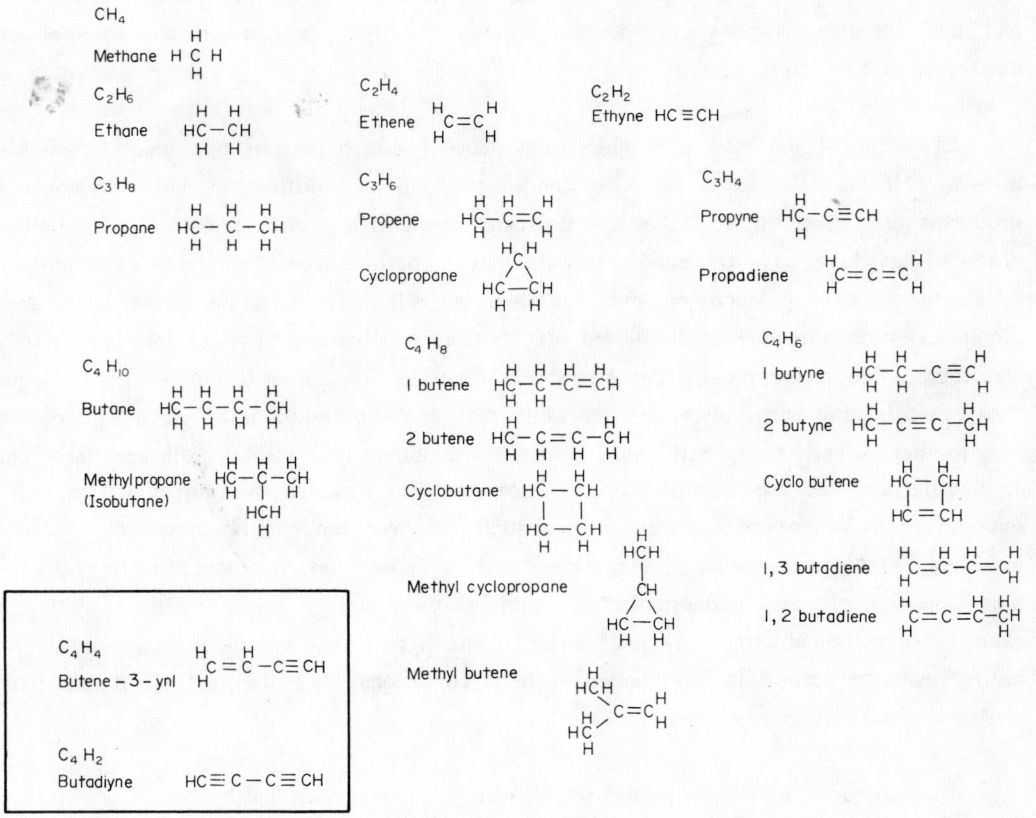

CH_4

Methane   H C H

C_2H_6

Ethane   HC—CH

C_2H_4

Ethene   H C=C H

C_2H_2

Ethyne   HC≡CH

C_3H_8

Propane   HC—C—CH

C_3H_6

Propene   HC—C=C

Cyclopropane   HC—CH

C_3H_4

Propyne   HC—C≡CH

Propadiene   C=C=C

C_4H_{10}

Butane   HC—C—C—CH

Methylpropane (Isobutane)   HC—C—CH

C_4H_8

1 butene   HC—C—C=CH

2 butene   HC—C=C—CH

Cyclobutane   HC—CH

Methyl cyclopropane   HC—CH

Methyl butene

C_4H_6

1 butyne   HC—C—C≡C

2 butyne   HC—C≡C—CH

Cyclo butene   HC—CH   HC=CH

1,3 butadiene   C=C—C=C

1,2 butadiene   C=C=C—CH

C_4H_4

Butene-3-ynl   H C=C—C≡CH

C_4H_2

Butadiyne   HC≡C—C≡CH

Figure 2-1.   A Very Small Sample of the Diversity Possible with Hydrogen-Carbon Compounds.

properties are known number. Figure 2-1 shows only a few common compounds, those possible with only carbon and hydrogen for carbon chains up to four carbon atoms long. As can be seen from the figure, the number of different isomers increases rapidly when the number of carbon atoms increases. For carbon chains 20 carbon atoms long, the number of isomers is in the thousands, and simple hydrocarbons with chains as long as 90 carbon atoms have been characterized. Add to this the possibility of replacing any or all of the hydrogen atoms with groups such as hydroxyl (OH), amino (NH_2), oxygen (O), sulfate (SO_3H), mercaptain (SH), cyano (CN), chlorine (Cl), fluorine (F), bromine (Br), and iodine (I), to name a few, and the potential diversity of this class of compounds becomes somewhat perceptible. The Handbook of Physics and Chemistry, published by the Chemical Rubber Company, lists over 15,000 organic compounds with some of their standard properties. Among these thousands of compounds are vitamins, drugs, and essential amino acids, as well as plastics, paints, lubricants, finishes, and so on. This all serves to illustrate the key and critical nature of hydrogen and carbon and their compounds, not only to the existence of our civilization but to the very existence of life itself.

Examples of the reversible absorption of occlusion of hydrogen by metals are well known. Metals used as cathodes in the electrolysis of solutions in which hydrogen is produced are observed to take up remarkable quantities of hydrogen. A palladium cathode may take up a thousand times its volume of hydrogen during the electrolysis of a water solution. Much of this hydrogen is released, in some cases with great violence, when the current is turned off, but a significant portion of it is retained in the metal. Electrochemically deposited iron may contain large quantities of hydrogen. Many metals absorb hydrogen directly when they are exposed to it at elevated pressures and temperatures. Iron, palladium, platinum, niobium, and nickel foils are permeable to hydrogen at elevated temperatures. Tantalum and titanium become embrittled when exposed to hydrogen under conditions in which the hydrogen can be absorbed. There is evidence that those metals that are most efficient as hydrogenation catalysts-- platinum, nickel, and palladium, for example--owe their activity to their ability to take up or combine with hydrogen. Much of the hydrogen absorbed in these metals can be removed by strong heating under vacuum conditions, but the last small quantities are often very difficult to remove.

The products of these processes in which metals take up hydrogen are termed "metallic hydrides." This class of hydrogen products (compound is too strong a term because of their instability and nonstoichiometric composition) is formed from a wide variety of metals and alloys with widely differing characteristics, so it is not surprising that their characteristics vary greatly. Because of this variety, it is difficult to state a set of characteristics common to all of the metallic hydrides, but some common trends can be generalized. In most cases, the properties of the hydrides are "metallic" and not markedly different from those of the parent metal or alloy. They possess appreciable electrical conductivity, their thermal conductivity is more metallic than nonmetallic, and they appear metallic in color and reflectivity. The amount of hydro- gen often bears no direct stoichiometric relationship, and compositions such as $MH_{1.3}$ or $MH_{.72}$ (where M is the metal) are common. These formulations are probably no more than reflections of the conditions under which the hydrogen saturation took place. In many cases, the absorption of hydrogen results in no fundamental changes in the metal crystal lattices, although density and X-ray determination do show the lattices to be expanded and perhaps slightly distorted. The available evidence seems to indicate that the hydrogen has entered the metal lattices, is occupying holes (interstices) be- tween the metal atoms, and is held in solid solution as with an alloy. Because of this structure, the terms "interstitial solid solution" and "interstitial compounds" are often applied to the metallic hydrides. The strong reducing properties of these hydrides, coupled with the metallic properties, led to the inference that the hydrogen is present in the atomic state, molecular hydrogen having undergone dissociation on entering the metallic lattice. These variations of properties among the hydrides led to a classification of all the noncovalent hydrides, as shown in Table 2-4. This classification helps to clarify the spectrum of properties found in the hydrides, but it cannot be taken as an absolute classification.

There has been much interest in and many investigations of the firmly bound and the loosely bound hydrides in the last 10 years, with the hope that formulations of hydrides could be found that would store hydrogen much like a sponge stores water. These hydrides could then be used as storage mediums for containing hydrogen for fuel-use applications and for transporting the gas from the production site to the user. A very large number of metals and metal alloys have been investigated, and several have been found that seem to have value in limited applications. The requirements placed on the performance of the materials in hydrogen storage are derived from the competitive methods of storage as-low pressure gas, storage as high-presure gas, and storage as a liquid. Low-presure storage is characterized by extreme bulkiness. Hydrogen, at room temperature and pressure, weighs only 0.081 gram per liter, versus air at 1.22 grams per liter and methane (natural gas) at 0.64 gram per liter. Storage of 1 kilogram of hydrogen with an energy equivalent equal to 146 megajoules would require a storage volume of 12,350 liters. The storage of the same amount of energy potential would require 2.88 kilograms of gasoline and would occupy a volume of only 4.1 liters. Increasing the pressure, of course, decreases the bulk of the stored hydrogen dramatically. The highest pressure commonly used in the shipment of compressed gas is about 400 atmospheres. Containment of the referenced 1 kilogram of hydrogen at a pressure of 400 atmospheres would reduce the volume required to 30 liters, a vast improvement but still much more volume than the same amount of energy in the form of gasoline. High-pressure gas storage always has the potential for leakage of the gas, should any part of the confinement system fail, and for outright explosion if the containment vessel should rupture.

Liquid hydrogen is an improvement over high-pressure gas. The referenced kilogram of hydrogen as liquid would occupy only 14.3 liters, still bulky but much less so than the gaseous storage.

Researchers in hydrides hoped to find a hydride that would improve on the density of the alternative hydrogen storage methods and that could be easily charged and dis-

Table 2-4. Hydride Classification

Hydride type	Saline hydrides	Semi-metallic hydrides	Firmly bound interstitial metallic hydrides	Loosely bound interstitial metallic hydrides	Absorption products
Examples	LiH NaH MgH_2	TiH_2 ZrH_2	PdH_x $LaN_{15}H_y$ $TiFeH_2$	MoH_x WH_y	Many substances
Characteristics	Stoichiometric composition. Salt like appearance. High decomposition temperature. Electric and thermal insulators	Near stoichiometric composition. Metallic appearance. High decomposition temperature. Poor electrical and thermal conductors	Variable composition dependent on the pressure temperature. History Metallic appearance. Modest decomposition temperatures. Fair electrical conductors	Variable composition dependent on the temperature. History Low hydrogen content Low decomposition temperature. Metallic properties	Hydrogen simply adsorbed on the surface

charged.   These researchers have to a large extent been successful in producing
hydrides that have a high volumetric efficiency for the storage of hydrogen.   Table 2-5
shows the volumetric figure of merit for the three advanced hydrides that show the
most promise as hydrogen storage mediums.   They absorb hydrogen when placed under
pressure with the evolution of modest amounts of heat, and when the pressure is
dropped, the hydrogen is desorbed with the absorption of the same modest amount of
heat that was evolved when the hydrogen was absorbed.   Figure 2-2 shows typical
desorption curves for hydrides.   This research was successful in discovering hydrides
that had greater hydrogen densities than present in liquid hydrogen, but there is doubt
whether they will ever be of commercial significance in the storage of hydrogen.   This
matter will be discussed at greater length in the chapter on hydrogen use.

The metallic hydrides have many properties in common with the base metals from
which they were formed.   The property that they do not share to any extent with the
metals is tensile strength and ductility.   Most of the hydrides are characterized by a
high degree of brittleness and very low physical strength.   With the advanced hydride
storage alloys, the absorption of hydrogen is accompanied by the crumbling of the
metal and the reduction of the pieces to rather small granules.   Cyclic addition and
removal of hydrogen causes this attrition process to continue until the original hydride
is reduced to nearly a dust-like consistency and particle size.   The absorption of the
hydrogen is also accompanied by a swelling of the solid material.   For example, the
fully hydrided FeTi alloy has a density of 5.47 grams per cubic centimeter (Table 2-5),
and the unhydrided alloy has a denisty of 6.49 grams per cubic centimeter.   When the
hydrogen is absorbed and desorbed, the volume of the solid changes about 11 percent.
These properties of the metallic hydrides create problems in the use of the hydrides as
storage mediums and when traces of them are formed in structural materials that are
used in the handling of hydrogen.   Both issues will be discussed further in Chapter 3.
More specific information on metallic hydrides is presented in Hydrides for Energy
Storage.[10]

Hydrogen gas has several properties that distinguish it from other gases.   It has
the lowest molecular weight of any gas, which results in hydrogen's having properties
that are the most extreme of all the gases.   It has the highest thermal conductivity,
velocity of sound, and mean molecular velocity and the lowest viscosity and density.
These properties are listed in Table 2-6 as hydrogen compares to air and helium.

These properties lead hydrogen to have a leak rate through small orifices faster
than other gases.   Hydrogen leaks about 1.4 times faster than helium, 2.8 times faster
than methane, and 3.8 times faster than air.   Although these differences are real, they

Table 2-5.   Advanced Metallic Hydrides Developed for Hydrogen Storage

Metal alloy	Alloy density	Hydride	Hydride density gram/cc	Wt % H_2	Hydrogen density litres/kilogram H_2
FeTi	6.49	$FeTiH_{1.9}$	5.47	1.80	10.15
$LaNi_5$		$LaNi_5H_{6.7}$	6.59	1.53	9.92
Mg_2Ni		Mg_2NiH_4	2.57	3.59	10.82
Reference	Gaseous hydrogen at 400 atmospheres pressure 30 litres Liquid hydrogen at 20.39°K 14.3 litres				

Hydride data taken from A. F. Andresen and A. J. Maeland, Editors *Hydrides for Energy Storage*, Pergamon Press, 1978.

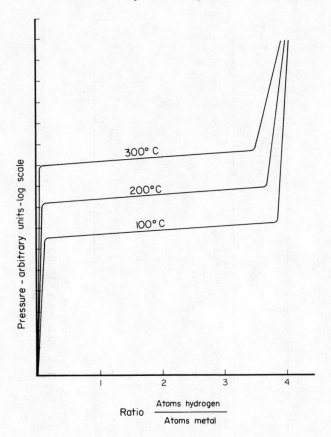

Figure 2-2. Generalized Representative Curves of Metallic
             Hydride Behavior.

are not so extreme that the leakage problem with hydrogen is signficantly different than that encountered with other gases.

The volumetric stoichiometric ratio for combustion of hydrogen with oxygen is two volumes of hydrogen to one volume of oxygen, which leads to a mass ratio of 0.125 unit weight of hydrogen to 1 unit of oxygen. Air is 20.946 percent oxygen, giving a volumetric stoichiometric combustion mixture ratio of 0.4189 volume of hydrogen to 1 volume of air and a mass ratio of 0.0292 unit weight of hydrogen to 1 unit weight of air. Converted to percentages, the stoichiometric mixture of hydrogen-oxygen contains 66.66 percent hydrogen by volume and 11.11 percent by weight; with air, the mixture contains 29.52 percent hydrogen by volume and 2.84 percent by weight. The stoichiometric mixture provides the maximum energy release per unit of matter and the

Table 2-6. Comparison of Gas Properties

Property	Units	Hydrogen	Helium	Air
Thermal conductivity (0°C)	cal/(sec)(cm^2)(°C/m) x 10^{-6}	388.46	324.00	54.22
Velocity of sound	m/sec	1284	965	331
Mean molecular velocity (0°C)	m/sec	1755	1252	463
Viscosity (20°C)	micropoise	87.6	194.1	184.1
Density (0°C)	grams/litre	.0899	0.178	1.293

Handbook of Physics and Chemistry
59 Edition, The Chemical Rubber Co. 1978.

highest flame temperatures, but of course other mixtures of hydrogen with oxygen and air will burn. Table 2-7 shows the flammability limits of hydrogen and methane with oxygen and air. As seen in the table, the flammability range for hydrogen is much broader than that of methane, but this increase in limits is almost entirely on the hydrogen-rich end of the range; the lower limit of hydrogen is about the same as that of methane. The high speed of sound in hydrogen and the high mean molecular velocity both contribute to a high flame velocity of hydrogen burning in air, about four times higher than the flame speed of methane in air.

Table 2-7. Limits of Flammability of Hydrogen and Methane with Percent
Air and Oxygen on a Volume Basis

	Oxygen		Air	
	Lower	Upper	Lower	Upper
Hydrogen	4.65	93.9	4.00	74.20
Methane	5.40	59.2	5.00	15.00

The product of combustion of hydrogen, in a simplistic sense is only water. Unfortunately, the equilibrium established at high temperatures contains small quantities of other chemical products. In Chapter 6, Figures 6-1 and 6-2 show the equilibrium products of combustion of hydrogen and gasoline at various mixture ratios. When the two substances are reacted in a fuel-lean mixture ratio, oxides of nitrogen are produced in quantities sufficient to be of concern as atmospheric pollutants. At exact stoichiometry, hydrogen produces a minimum of nonwater products (all less than 0.1 part per million by weight); gasoline, however, shows significant carbon monoxide at exact stoichiometry. As the mixture ratio is shifted into the fuel-rich regime, the oxides of nitrogen are no longer produced, but unburned hydrogen and trace amounts of ammonia are observed in both hydrogen and gasoline combustion. In addition to these products, which will occur whenever any fuel that contains hydrogen is combusted with air, all the products of fuel-rich combustion of carbon are present. These include unburned hydrocarbons and carbon monoxide. Hydrogen combustion produces a somewhat higher temperature than the combustion of gasoline and, as a result, the quantity of nitrogen oxides peaks at somewhat higher concentrations.

Liquid hydrogen is a colorless fluid that, when confined in a properly insulating vessel, looks much like water. Its main significant property is its extreme cold, only

20.3°K above absolute zero. It can be stored in containers, pumped through pipelines, metered, and restricted by valves of ordinary design. It has no effect, other than temperature, on any substance and does not react chemically in any manner. The liquid hydrogen of commerce has all been converted by catalytic treatment to the equilibrium mixture of para and ortho hydrogen. This mixture contains more than 99 percent para hydrogen, so the properties quoted are those of para hydrogen. Table 2-8 lists the known properties of liquid and gaseous hydrogen.

Table 2-8. Properties of Hydrogen

Molecular weight, amu	2.016
Triple point pressure, atm	0.0695
Triple point temperature, °K	13.803
Normal boiling point (NBP) temperature, °K	20.268
Critical pressure, atm	12.759
Critical temperature, °K	32.976
Density at critical point, g/cm^3	0.0314
Density of liquid at triple point, g/cm^3	0.0770
Density of solid at triple point, g/cm^3	0.0865
Density of vapor at triple point, g/m^3	125.597
Density of liquid at NBP, g/cm^3	0.0708
Density of vapor at NBP	0.00134
Density of gas at normal temperature and pressure (NTP), g/m^3	83.764
Density ratio: NBP liquid-to-NTP gas	845
Heat of fusion, J/g	58.23
Heat of vaporization, J/g	445.59
Heat of sublimation, J/g	507.39
Heat of combustion, kJ/g to steam at 100°C	119.93
Heat of combustion, kJ/g to water at 0°C	141.86
Specific heat (CP) of NTP gas J/g·°K	14.89
C_p of NBP liquid, J/g·°K	9.69
Specific heat ratio (C_p/cu) of NTP gas	1.383
Specific heat ratio (C_p/cu) of NBP liquid	1.688
Viscosity of NTP gas, g/cm·s	8.75×10^{-5}
Viscosity of NBP liquid, g/cm·s	1.33×10^{-4}
Thermal conductivity of NTP gas, mW/cm·°K	1.897
Thermal conductivity of NBP liquid, mW/cm·°K	1.00
Surface tension of NBP liquid, N/m	1.93×10^{-3}

Dielectric constant of NTP gas	1.00026
Dielectric constant of NBP liquid	1.233
Index of refraction of NTP gas	1.00012
Index of refraction of NBP liquid	1.110
Adiabatic sound velocity in NTP gas, m/s	1294
Adiabatic sound velocity in NBP liquid, m/s	1093
Compressibility factor (z) in NTP gas	1.0006
Compressibility factor (z) in NBP liquid	1.712×10^{-2}
Gas constant (R), $cm^3 \cdot atm/g \cdot °K$	40.7037
Isothermal bulk modulus (α) of NBP liquid MN/m^2	50.13
Volume expansivity (β) of NBP liquid, $°K$	1.658×10^{-2}
Limits of flammability in air, vol. %	4.0 to 75
Limits of detonability in air, vol. %	18.3 to 59.0
Stoichiometric composition in air, vol. %	29.53
Minimum energy for ignition in air, mJ	$2. \times 10^{-2}$
Autoignition temperature, $°K$	858
Hot air-jet ignition temperature, $°K$	943
Flame temperature in air, $°K$	2318
Thermal energy radiated from flame, %	17 to 25
Burning velocity in NTP air, cm/sec	265 to 325
Detonation velocity in NTP air, km/sec	1.48 to 2.15
Diffusion coefficient in NTP air, cm^2/s	0.61
Diffusion velocity in NTP air, cm/sec	~2
Buoyant velocity in NTP air, m/sec	1.2-9
Maximum experimental safe gap in NTP air, cm	8×10^{-3}
Quenching gap in NTP air, cm	6.4×10^{-2}
Detonation induction distance in NTP air, L/D	~100
Limiting oxygen index, vol. %	5.0
Vaporization rates of liquid pools, cm/min	2.5 to 5.
Burning rate of spilled liquid pools, cm/min	3.0 to 6.6

NBP = Normal boiling point
NTP = Normal temperature and pressure = 1 atm, 20°C

Properties are for equilibrium parahydrogen-orthohydrogen at the temperature at which the property was measured.

Data adapted from the U.S. Department of Commerce/National Bureau of Standards Technical Note 690, J. Hord, October 1976.

Chapter 2 References

1.   E.W. Washburn and H.C. Urey, Proc. Natl. Acad. Sci., Vol. 18, p. 496, 1932.

2.   T. Moeller, Inorganic Chemistry, John Wiley and Sons, Inc., New York, 1952.

3.   H.C. McIntyre, "Natural-Uranium Heavy-Water Reactors," Scientific American, Vol. 233, No. 4, p. 17, October 1975.

4.   W. Heisenberg, Z. Physik, Vol. 41, p. 239, 1937.

5.   C.A. Bailey, editor, Advanced Cryogenics, Plenum Press, London and New York, 1971.

6.   E.B. Maxted, Modern Advances in Inorganic Chemistry, Clarendon Press, Oxford, 1947.

7.   A. Stock, Hydrides of Boron and Silicon, Cornell University Press, Ithaca, New York, 1933.

8.   F. Paneth, Radio Elements as Indicators and Other Selected Topics in Inorganic Chemistry, McGraw-Hill Book Company, New York, 1928.

9.   H.C. Brown, Organic Synthesis via Boranes, John Wiley and Sons, New York, 1975.

10.   A.F. Andresen and A.J. Maeland, editors, Hydrides for Energy Storage, Pergamon Press, 1978.

Chapter 3

CURRENT INDUSTRIAL USES
OF HYDROGEN

The amount of hydrogen used industrially is quite large, with the industries using the hydrogen usually manufacturing it and using it onsite. This results in hydrogen's failing to show up on the prepared list of industrial commodities, and accounting for the total amount produced is complex. Table 3-1 gives one estimate of the amount of hydrogen used by the major current technologies consuming hydrogen.[1] Each of these industries produces and consumes hydrogen in a somewhat different manner, and each will be discussed in turn to show how it fits into the total hydrogen use structure.

In the petrochemical industry, hydrogen is both a product and a raw material in the processes used to convert a barrel of crude oil into the petroleum products of commerce. The details of these processes vary greatly, depending both on the type of crude oil being processed and on the design of the refinery, but a common thread exists for all these processes that can demonstrate how the hydrogen is produced and used.

In the first step, the crude oil is heated sequentially to higher and higher temperatures. As this process starts, the most volatile components begin to boil off and the most unstable begin to decompose. The decomposition of unstable components can produce hydrogen gas. The vapor fraction of this step, containing some methane, ethane, propane, butane, and hydrogen, is collected, and the individual substances are separated for later uses. As the heating continues, compounds that have higher and higher boiling points are removed as vapors, and more stable compounds decompose to produce hydrogen. This process is continued until the final remaining residue is subjected to temperatures in the range of 400° to 500°C. The resulting end material is a coke-like black substance that contains no volatile materials and is almost free of hydrogen. The vapors that were produced during the heating process are collected by condensation to provide the various petrochemicals.

Table 3-1. Hydrogen Supply-Demand Relationships, 1964-75

(Billion standard cubic feet)

	1964	1965	1966	1967	1968	1969	1970	1971	1972	1973	1974	1975°
World production												
United States	990	1,200	1,440	1,650	1,870	2,030	2,170	2,305	2,480	2,575	2,570	2,515
Rest of the world	1,500	1,575	2,000	2,420	2,995	3,210	3,940	4,180	4,600	4,800	5,050	5,145
Total	2,490	2,775	3,440	4,070	4,865	5,240	6,110	6,485	7,080	7,375	7,620	7,660
Components of U.S. supply:												
Production	990	1,200	1,440	1,650	1,870	2,030	2,170	2,305	2,480	2,575	2,570	2,515
Distribution of U.S. supply:												
Demand	990	1,200	1,440	1,650	1,870	2,030	2,170	2,305	2,480	2,575	2,570	2,515
U.S. demand pattern:												
Petroleum refining	290	375	475	640	775	850	900	995	1,045	1,105	1,065	1,080
Chemicals:												
Ammonia	496	596	714	748	795	867	936	983	1,027	1,020	1,060	1,075
Methanol	95	103	118	124	137	151	177	177	233	254	247	180
Cyclohexane	22	27	30	28	32	35	29	26	36	34	37	29
Benzene	13	15	17	17	17	10	7	6	12	20	21	15
Hexa-methylene-diamine	—	—	—	—	12	12	12	13	15	17	17	15
Aniline	2	3	3	3	4	4	5	4	5	6	7	5
Naphthalene	8	10	11	11	11	11	9	8	7	7	6	7
Oxo alcohols	4	4	5	6	6	6	6	6	7	7	7	6
Other chemicals	10	12	12	13	16	14	19	17	18	25	23	23
Total chemicals	650	770	910	950	1,030	1,110	1,200	1,240	1,360	1,390	1,425	1,355
Miscellaneous uses:												
Hydrogenation of oils	7	7	7	7	7	7	7	8	9	10	10	10
Liquid hydrogen	18	21	22	22	22	21	21	17	17	14	14	14
Direct reduction	—	—	—	—	—	5	6	11	12	17	18	18
Other	25	27	26	31	36	37	36	34	37	39	38	38
Total	50	55	55	60	65	70	70	70	75	80	80	80
Total U.S. demand	990	1,200	1,440	1,650	1,870	2,030	2,170	2,305	2,480	2,575	2,570	2,515

°Estimated

Taken from Mineral Facts and Problems,
U.S. Bureau of Mines Bulletin 667, 1975.

The processes used to treat the collected products to produce specific substances also generate and use hydrogen. An example of one of the hydrogen-producing processes is the very important manufacture of ethylene. Ethylene is used in large quantities as the feedstock for the production of ethylene glycol used as antifreeze in automobiles, and it is also used in very large quantities (1.9×10^9 kilograms in 1979) for the production of polyethylene.

Some ethylene (C_2H_4) is obtained directly during the initial fractionation of the crude oil; in addition, a significant quantity of ethane (C_2H_6) is obtained. Ethane can be converted to ethylene by subjecting it to a high temperature in the presence of a catalyst. The reaction is as follows:

$$CH_3 - CH_3 \quad \text{Heat} \longrightarrow CH_2 = CH_2 + H_2$$
$$\text{Catalyst}$$

The ethylene is separated from the hydrogen by liquefaction, and the hydrogen is diverted to other processes.

A petroleum refining process that uses large quantities of hydrogen is the catalytic cracking process used for the production of gasoline. In the initial boiling process, the amount of gasoline obtained ranges from as low as 15 percent to as much as 50 percent, depending on the crude oil composition. Much of the other material obtained can be converted to gasoline by using the catalytic cracking process. In crude oil, the compounds have chains of carbon atoms of various lengths; in gasoline, the chain length is in the range 5 to 10 carbon atoms. The higher boiling fractions of the crude oil have carbon chain lengths ranging from 10 carbons to as high as 50 or 60, with complex branching and ring structures. In gasoline, the ratio of hydrogen to carbon ranges from 2.4 to 2.2, with the higher boiling fractions always less than 2.2. If the higher boiling fractions of the crude oil are reheated to high temperatures in the presence of catalysts, the long chains of carbon atoms will break down into the shorter chain lengths desirable for the production of gasoline. When they break down, they still maintain their low ratio of hydrogen to carbon. If they were used in this hydrogen-deficient form, the gasoline would not perform well in an automobile and would be quite unstable in storage. To convert the initial hydrogen-deficient output from the chain-breaking processes to quality gasoline, it is subjected to contact with hydrogen under high pressure in the presence of a catalyst. The hydrogen reacts with the hydrogen-deficient chains and produces a product that is difficult to distinguish from the gasoline obtained by direct vaporization of the crude oil.

The processes used to reduce the sulfur content of petroleum products also use hydrogen. Crude oil contains sulfur combined in several different types of compounds: In mercaptans, the chemical linkage is the same as that in alcohols, except that the oxygen atom of the alcohol is replaced with a sulfur atom:

$$-CH_2-OH \text{ (alcohol)} \qquad\qquad -CH_2-SH \text{ (mercaptan)}$$

In thio-ethers, the sulfur replaces oxygen in the ether linkage:

$$-CH_2-O-CH_2- \text{ (ether)} \qquad\qquad -CH_2-S-CH_2- \text{ (thio-ether)}$$

In sulfides, the sulfur is combined with a metal and, in sulfates, oxidized sulfur is combined with a metal. The diversity of these compounds and their diverse chemical reactivity make their removal difficult in a single chemical process. Rather than use a separate process for the removal of each of these compounds, the bulk crude oil, or the separated component, is subjected to hydrogen at elevated temperature and pressure and virtually all the sulfur compounds are converted to hydrogen sulfide:

$$-S- \; + H_2 \longrightarrow H_2S$$

Hydrogen sulfide is a gas at room temperature and is a weak acid. Its gaseous nature causes it to escape from the oil product, and its acid nature allows it to react with a strongly basic substance such as lime (calcium oxide) or limestone (calcium carbonate), to give stable calcium sulfide. This process removes the bulk of the sulfur from the crude oil product, giving a product that is more stable in use and storage, and one that is much more environmentally acceptable in use. So it can be seen that hydrogen is absolutely essential for the production of high-grade clean fuels from crude oil. As the cleaner grades of crude oil are used up and it becomes necessary to use the less desirable, heavier, higher sulfur oils, the demand for hydrogen will increase.

Gasification and liquefaction of coal for the production of more desirable fuels will require more hydrogen than the upgrading of crude oil. Coal, like crude oil, is not a uniform substance. Large variations have been observed in the composition of the various types of coal, within a coal type, and even within a particular coal field. These variations include the volatile fraction, amount of ash, decomposition temperature, amount of sulfur, amount of radioactive substances, and all other physical and chemical properties. All coals, however, possess significant amounts of free carbon and hydrocarbon compounds that can be combusted directly to produce energy or can be converted through chemical processes to other hydrocarbon fuels that can be used like oil products.

To understand how the use of hydrogen is necessary for the upgrading of coal to better quality fuels, it is useful to consider coal, a very low grade crude oil, having a melting point well above room temperature (which means it must be handled as a solid) and a hydrogen-to-carbon ratio of less than one, often much less than one. To upgrade coal to produce more desirable fuels, the addition of even more hydrogen is necessary than was necessary to upgrade crude oil. The upgrading of crude oil required that the hydrogen-to-carbon ratio be brought from between 1 and 2 to about 2.3 or, on a weight basis, about 1 kilogram of hydrogen for every 15 kilograms of crude oil. To ugrade coal from a hydrogen ratio of 0.1 will require about 2 kilograms of hydrogen per 15 kilograms of coal. Although it requires more hydrogen to upgrade coal to a higher quality fuel, the process is much the same. The coal is ground to a fine powder to allow good contact of the hydrogen with the carbon. The fine coal is placed in a reactor with hydrogen at high pressure, and the mixture is heated to high temperatures in the presence of a catalyst. The mixture of compounds in the coal reacts with the hydrogen to produce hydrocarbon compounds of the types present in crude oil. The higher the temperature, the more of the volatile compounds used in gasoline are formed. At the same time that the carbon compounds are being converted to hydrocarbons, much of the sulfur is being converted to "sour gas," hydrogen sulfide, which can be removed from the final product with the lime or limestone.

As described in Chapter 2, hydrogen can be produced from coal and water. In a plant designed to produce liquid fuels from coal, the following steps can be outlined: First, part of the coal is reacted by one of the various processes to produce hydrogen. Second, the hydrogen is purified as necessary and reacted with more coal to produce the liquid fuels. A significant amount of ash from the coal and calcium sulfide from the sulfur are also produced by this process and must be discarded.

Currently, the largest and most important single use of hydrogen is in the production of ammonia by the Haber process. Ammonia and its derivatives have wide use in industry and commerce in products ranging from plastics to cleaning agents, but the main and most critical use is as fertilizer. This use of hydrogen has probably made a bigger contribution to the viability of modern man and the growth of world population than any other single technological development. Without the ability to combine hydrogen with nitrogen to produce ammonia for fertilizer, it would not be possible to feed the current population of the world.

Around 1900, all of the nitrogenous material used for fertilizer came from salt deposits, such as those in Chile, from animal waste, or from that nitrogen fixed by the bacteria living in symbiosis with one type of plant, the legumes. The nitrates from

Chile were being depleted at an alarming rate, as were the sources of animal wastes that could be profitably mined--bird dung from tropical ocean islands and bat dung (guano) from caves. It was recognized that new sources of fixed nitrogen must be found if we were to be able to feed ourselves. A 15-year research effort, with participants from all over the world, finally resulted in the discovery of the Haber process for the synthesis of ammonia.

In this process, nitrogen from air is reacted with hydrogen to produce ammonia:

$$N_2 + 3H_2 \longrightarrow 2\ NH_3$$

This seemingly simple reaction is, in fact, very difficult to perform. The difficulty arises from several factors, the most important of which is the high stability of the nitrogen molecule and the absence of any stable intermediate compounds. The high stability of the nitrogen molecule prevents it from reacting with anything at modest temperatures, and temperatures high enough to activate it are so high that the desired ammonia molecule is not stable. The absence of any stable intermediate compounds meant that it was necessary to get the one nitrogen and three hydrogen molecules to come together at once, something very improbable in a reacting gas mixture. The Haber process works because a catalyst was found that would activate both the nitrogen and hydrogen molecules to react at a temperature low enough so that the resultant ammonia is stable. It is necessary, however, to perform this reaction at very high pressures, to push close together as many hydrogen and nitrogen molecules as possible.[2]

Hydrogen is also used by the edible oil industry. Many natural edible oils have a large amount of unstaturation, which in this application is defined as potential sites within the molecule that could be combined with hydrogen but are not. If the oils have sufficient unsaturation, linseed oil or tung seed oil for example, the oil is a drying oil. The reactive sites within the molecule that might combine with hydrogen react instead with oxygen of the air to crosslink the oil molecules into a plastic-like infusable solid. This property is exploited in the use of these oils in the manufacture of paints; for use as food oil, however, this tendency to form solids is very undesirable, because the solid formed is no longer a food. Even if the oils contain less unsaturation than the drying oils, they are still reactive with oxygen to form compounds that promote the growth of bacteria, so they are not very stable or storable. These unsaturated oils can be treated with hydrogen to fill all or most of the reactive sites that can combine with the hydrogen. The resultant oil, termed partially or fully hydrogenated, is much more stable to storage and the growth of bacteria. The hydrogenation usually results in raising the melting point of the oil, such that an oil that had been liquid at room temperture becomes a solid when hydrogenated.

The hydrogen required for this use must be very pure, for the resultant product will be a food material. In some instances, the hydrogen is produced by electrolysis because that is the most pure type of hydrogen commonly available; in other cases, the hydrogen is produced by other processes and is then subjected to a stringent purification cycle. The hydrogenation process, like those used to add hydrogen to crude oil and coal, uses high temperatures, catalysts, and high hydrogen pressure. A common catalyst for this process is Rainey nickel, a highly active form of metallic nickel that activates the hydrogen to react rapidly with the unsaturated sites in the oil.

Hydrogen is a critical element in the production of polyurethane plastics. It is required in the production of toluene diisocyanate (TDI), the most important active ingredient in the formulation of these plastics. TDI is synthesized from toluene in the following manner:

$$\text{(1)} \quad \text{(2)} \qquad\qquad \text{(3)}$$
$$C_7H_8 + 2HNO_3 \longrightarrow C_7H_6(NO_2)_2 + 2H_2O$$
$$\text{(3)} \qquad\qquad \text{(4)}$$
$$C_7H_6(NO_2)_2 + 6H_2 \longrightarrow C_7H_6(NH_2)_2 + 4H_2O$$
$$\text{(4)} \qquad \text{(5)} \qquad\qquad \text{(6)}$$
$$C_7H_6(NH_2)_2 + 2COCl_2 \longrightarrow \underline{C_7H_6(NCO)_2} + 4HCl$$

Toluene (1) is a product extracted from crude oil or coal tar. It is treated with nitric acid (2) in the presence of concentrated sulfuric acid as a dehydrating agent to form dinitrotoluene (3). The dinitrotoleune (3) is washed free of acids and exposed to hydrogen at very high pressure in the presence of a catalyst. The hydrogen reacts with the nitro groups to form water and diaminotoluene (4). The diaminotoluene (4) is then reacted with carbonyl chloride (5) (phosgene) to form TDI (6). One kilogram of hydrogen is required for the manufacture of 14.5 kilograms of TDI.

TDI is reacted with long chain dialcohols in proportions adjusted to allow one TDI to attach to one alcohol group, using only one of the isocyanate groups:

$$2\ C_7H_6(NCO)_2 + HO-R-OH \longrightarrow \underline{OCN-C_7H_6NH-R-NHC_7H_6NCO} + 2CO_2$$

where R = a 4- to 10-carbon atom chain, fully saturated. The underlined compound is the basic structure of the urethane-type plastics. It is further reacted with dialcohols of the same or different chain length to provide the fully polymerized structure. The carbon dioxide released when the TDI reacts is the foaming agent that causes the urethane to expand to the foamed state.

Hydrogen is also involved in the production of methanol (synthetic wood alcohol), a process in which the hydrogen need not be separated from the synthesis gas stream in which it is produced. The reaction for this process is

$$CO + 2H_2 \longrightarrow CH_3OH$$

Some of the methods of producing hydrogen from fossil fuels produce a gas mixture approximating the starting composition for the synthesis of methanol; these methods will be discussed in the next chapter. For the commercial production of methanol, one of the several processes that produce a mixture of hydrogen and carbon monoxide is selected.   The gas from this reaction is cleaned to remove the residual sulfur compounds present, and the mixture of hydrogen and carbon monoxide is passed over a precious metal catalyst where they react to form the methanol.   The reaction is favored by high temperatures and pressures.

Hydrogen is used in a number of processes in the metallurgical industry. It can be used as a reducing agent to react with the metal oxides to produce the free metal, or it can be used as an atmosphere for the protection of metals during high-temperature fabrication processes.

If iron ore (iron oxide) is treated with hydrogen at elevated temperatures, the hydrogen reacts with the oxygen of the iron oxide to form water, thus releasing the iron metal in a free state.   The iron reduced by this process is free of the usual impurities introduced when iron is reduced by coal. These common impurities--sulfur, phosphorous, silicon, and carbon--are undesirable in the formulation of certain special-purpose iron alloys.   For these applications, the iron produced by direct reduction by hydrogen is preferred.

In the manufacture of tungsten, the presence of carbon in the metal makes it extremely brittle. Tungsten containing carbon is so brittle that it is almost impossible to form it by ductile forming operations such as drawing, forging, stamping, and the like. For example, in the manufacture of light bulb filaments, the tungsten must be drawn through a series of dies to reduce it to very fine wire, a process that is nearly impossible with tungsten containing carbon. All the tungsten used for the production of filaments is produced by the direct reduction of tungsten oxide with hydrogen, and much of the tungsten used in other processes is produced in this manner.

In addition to its use in the direct reduction of metal oxides to the pure metal, hydrogen is used by the metallurgical industry for several other purposes. One of the more important of these is as a reducing blanket gas for high-temperature forming processes. In brazing operations involving the joining of high-temperature alloys, the parts to be brazed are coated on the surfaces to be joined with the brazing alloy and clamped in place. The assembled parts with their holding jigs are then placed in a furnace. The furnace is purged with hydrogen gas to remove the last traces of oxygen and then raised to a temperature somewhat higher than the melting point of the brazing alloy. During the heating operation, the hydrogen purge is maintained to prevent the leakage of oxygen back into the furnace and to reduce any oxide film on the surface of the metal that might interfere with the formation of a strong joint. At the high temperature, the brazing alloy melts and wets the adjoining surfaces of the metal parts being joined. Traces of oxide films on the surfaces inhibit wetting, and the presence of the high-temperature hydrogen reduces the oxide films and prevents their reformation. To form a strong bond between brazed parts it is necessary that both surfaces to be joined be completely wet by the molten brazing alloy. After sufficient time at a temperature that allows wetting to occur, a time usually determined by experiment, the parts, still in the hydrogen atmosphere, are cooled to room temperature. When cooled, the parts are removed from the hydrogen atmosphere for use.

Hydrogen has been, and will remain, the best fuel for rocket engines. Hydrogen combined with fluorine is the most energetic combination of two chemicals possible for a rocket propellant system. This combination in a rocket engine operating at a chamber pressure of 1000 pounds per square inch produces a specific impulse (the number of seconds that 1 kilogram of propellant can produce 1 kilogram of thrust) of 410 seconds. Fluorine, however, is extremely difficult to handle, very expensive, and very toxic, properties that have thus far prevented, and probably always will prevent, its use in an operational rocket system. Fortunately, the combination of hydrogen with oxygen produces a specific impulse of 390 seconds, only 5 percent less than that of fluorine and hydrogen. Oxygen is relatively inexpensive, easy to handle, and nontoxic, and its combination with hydrogen will continue to be the propellant of choice for rocket systems. Both of the upper stages of the Saturn V launch vehicle used in the Apollo missions to the moon and the Skylab launch used the combination of hydrogen and oxygen to provide the performance needed to make these missions possible. The upper stage that pushed the Viking mission to Mars and the Voyager mission to Saturn, the Centar, used hydrogen and oxygen as its propellants. Without the use of these propellants neither of these missions could have used the Titan as the first stage, but would have required a much larger, more costly launch vehicle.

Future space missions will all use hydrogen and oxygen as a mainstay of their propulsion systems,[3] for these are the main engine propellants in the Space Shuttle system. The Space Shuttle has two large, solid rockets that are part of the initial lift-off propulsion, but the main engines are fueled with liquid hydrogen and the oxidizer is liquid oxygen. Underneath the orbiter is a very large drop tank designed to hold liquid hydrogen and oxygen in separate compartments.   Large feedlines run from these tanks into the orbiter, where they are attached to the engines.   The orbiter also has small tanks of hydrogen and oxygen for the final insertion into orbit.   Many of the special spacecraft that the Shuttle will take into orbit over the next decade will also use the hydrogen and oxygen fuel combination, which will provide the best compromise of high performance and safe and easy handling on the launch pad.

The uses discussed thus far--petrochemical processing, ammonia synthesis, lique-faction of coal, hydrogenation of edible fats and oils, metallurigical processes, and rocket propellants--account for more than 90 percent of the hydrogen produced.   There are, however, a number of other uses that do not account for a large share of the total production but are so totally dependent on some physical or chemical property of hydrogen that there are no possibilities of substitutes.

For example, hydrogen is the essential component in the operation of the flame ionization detector used with the gas chromatograph, an instrument widely used to per-form chemical analyses.   In this instrument, an unknown sample of volatile material is swept through a long narrow tube (called a column) with a stream of helium gas.   The tube is filled with a granular material coated with a material that interacts weakly with the class of materials expected in the sample.   The coating material is selected to interact by adsorption, absorption, and solution, but not to react chemically.   As the sample is swept through the tube, the various components are differentially retarded by their interaction with the coating material.   At the tube outlet, the components are sufficiently separated that they exist individually.   There are a number of methods for detecting the components of the original sample as they come out of the separation process, but one of the most sensitive and widely used is the hydrogen flame ionization detector.

The hydrogen flame ionization detector is made up of a small flame, possibly ¼-inch long, of very pure hydrogen burning in purified air.   Small electrodes are placed in the flame, and a potential of 5 to 30 volts is placed across the electrodes.   The high stability of hydrogen, the water produced by the combustion, and the nitrogen from the air result in virtually no ionization of the gases in the flame and almost no current flow between the electrodes, despite a flame temperature of over 1000°C. Helium used as the carrier gas in the gas chromatograph is also very stable at this temperature, such that its presence produces no ionization.   When a component of the orginal sample that contains carbon exits from the tube, the high temperature burns the carbon compound and produces many intermediate species that produce ionization. This ionization allows a current to flow between the electrodes, which can be amplified and fed to a recorder to show the presence of the different substances in the hydrogen flame. This effect is extremely sensitive. With the hydrogen flame ionization detector, it is possible to detect a quantity of a carbon-containing substance as small as 10^{-12} grams (1 picogram).

By calibrating the time it takes a known substance to pass through the column and the strength of the electrical signal from a known amount of the known substance, it is possible to use the gas chromatograph to identify and quantify the substances in many types of unknown samples. It is used for the chemical analysis of materials as diverse as crude oil, petrochemicals, and biological fluids, both in research diagnosis and forensic investigations; for testing water for toxic substances such as pesticides; and for control of chemical processes. At this time in these very important and critical applications, there is no more simple or reliable a detector than the hydrogen flame ionization detector.

Hydrogen has the most simple atomic nucleus, one lone proton. Because of the simplicity of the nucleus, it makes an ideal target for studying the interaction of high-energy particles with nuclear materials. In these applications, hydrogen serves as both the target for the high-energy particles and the detection medium for the interactions. These detectors consist of a large container of liquid hydrogen held very near its boiling point. The chamber is equipped with a piston or some other means of quickly reducing the pressure on the liquid hydrogen. When the particles from the accelerator are directed into the chamber, they collide with the simple hydrogen nuclei and new particles are produced that fly off in all directions and carry information about the nature of the collision. At almost the same instant of time as the collision, the piston is quickly withdrawn, reducing the pressure on the liquid hydrogen. At the moment that the piston is withdrawn, the hydrogen should boil, but it takes some period of time for boiling to commence, and the boiling hydrogen bubbles need something on which to form. The fragments resulting from the collision cause ionization in the liquid hydrogen, and ions form perfect sites on which the incipient bubbles of boiling can form.

The result of all these actions is a thin stream of bubbles tracing the path of the fragments. A high-speed photograph is taken at the exact moment of the formation of the bubbles, and it contains information about the type of particles, their number, and their energy. These photographs are analyzed and used as a main source of data by the physicist in investigating the deep innner structure of matter. The liquid hydrogen bubble chamber has contributed a large portion of this data.

Hydrogen is still used as a lifting gas for balloons. In this application, it actually provides about 8-percent greater lift than does helium and is usually less costly. Helium must be transported in high-presure cylinders, but hydrogen forms simple solid chemical compounds that can easily be released. One of the hydrogen compounds that has been used for this purpose is lithium hydride (LiH), which reacts with water to release hydrogen in the following manner:

$$LiH + H_2O \longrightarrow LiOH + H_2$$

One kilogram of lithium hydride will produce 2800 liters (100 cubic feet) of hydrogen gas. Lithium hydride is a white powder that is completely stable as long as it is not allowed to contact reactive substances such as water. A small package of lithium hydride can be carried about for months without change and then added to a bottle of water to produce hydrogen. This technique is used to launch small weather balloons or deploy radio antennas in remote areas of the earth. Because of its flammability and the availability of helium, hydrogen is no longer used in large manned balloons or in balloons carrying valuable instruments.

Chapter 3 References

1.   D.C. Adkins and R.J. Jaske, "Hydrogen," Mineral Facts and Problems, U.S. Bureau of Mines Bulletin 667, 1975 edition.

2.   J.C. Bailar, Jr., editor, Comprehensive Inorganic Chemistry, Vol. 2, Pergamon Press, Oxford, 1973.

3.   W.H. Sterbentz and J.H. Guill, "Cryogenic Propulsion for Long Duration Space Mission," Proceedings of the International Symposium on Space Technology and Science, Tokyo, 1965.

General Review Sources

Hittman Associates, Inc., "Economics and Market Potential of Hydrogen Production," U.S. Department of Energy Contract 31-109-38-4409, Columbia, Maryland, 1978.

Kelly, J.H. and E.A. Laumann, "Hydrogen Tomorrow," Jet Propulsion Laboratory, Pasadena, California, 1975.

U.S. Department of Energy, "Supply and Demand of Hydrogen as a Chemical Feedstock," Workshop held at University of Houston, C.J. Huang, editor, 1977.

Chapter 4

METHODS OF PRODUCING HYDROGEN

SIMPLE LABORATORY METHODS

A number of simple methods can be used in the laboratory to produce small quantities of hydrogen. The most commonly used is a reaction of metal with acid. Any metal that is more reactive than hydrogen will react with acid solutions to produce hydrogen as follows:

$$Acid\ (H) + Metal \longrightarrow Hydrogen + Metal\ Salt$$

Simple acids like hydrochloric or sulphuric acid will react with metals such as iron and zinc in the following manner:

$$2HCl + Fe \longrightarrow H_2 + FeCl_2$$

$$H_2SO_4 + Zn \longrightarrow H_2 + ZnSO_4$$

These reactions are easily controlled and can be conducted in simple laboratory glassware.

Hydrogen can be produced by the reaction of steam with a number of metals at modestly elevated temperatures. In this method, iron is the most commonly used metal because the reaction is easily controlled. Magnesium and aluminum can also be used, but the action of these metals on steam is much more vigorous and hard to control. With iron, the reaction proceeds as follows:

$$2Fe + 3H_2O \longrightarrow 3H_2 + Fe_2O_3$$

49

Hydrogen can be generated by the action of very strong solutions of sodium hydroxide on aluminum, but this reaction is impractical in the production of significant quantities of hydrogen because of the cost of the reagents.   However, during the Second World War, because of an unusual set of circumstances, hydrogen produced in this manner was used to inflate small balloons to raise radio antennas and to observe weather conditions.   The airfields where the balloons were launched would often have a supply of damaged and scrapped aluminum aircraft parts.   The scrap parts would be cut into small pieces and placed in an ordinary 55-gallon steel drum, and appropriate valves would be installed on the drum for controlling the hydrogen flow and admitting the sodium hydroxide solution.   A 10- to 20-percent solution of sodium hydroxide (lye) in water would then be allowed to flow into the drum.   The aluminum scrap would dissolve, releasing hydrogen at a rate controlled by the rate of flow of the sodium hydroxide solution into the drum.   The hydrogen was taken directly from the drum to inflate the balloon.   The chemical reaction for this method of producing hydrogen is as follows:

$$Al + 3NaOH \longrightarrow Na_3AlO_3 + 1.5H_2$$

Hydrogen can be produced by the reaction of metallic hydrides with water or acids.   As mentioned in Chapter 2, a number of metals react directly with hydrogen at elevated temperatures to produce hydrides, and another group of metallic hydrides can be produced by indirect methods.   In the case of the salt-like hydrides of metals such as lithium, magnesium, and aluminum, the reaction with water is quite rapid, with aluminum and magnesium hydride almost explosively rapid.   The reaction of these hydrides with acids is explosively rapid and quite dangerous.   With the more inert hydrides titanium and zirconium, the reaction with water is relatively slow and may require some heating.   The reaction of these hydrides with acids is much more rapid but not of explosive violence.

These simple reactions for the production of hydrogen are discussed more for their chemical interest than as examples of methods used in any significant manner for the production of hydrogen.   They have all, however, been used under special conditions for the production of hydrogen for special purposes.

HYDROCARBONS

In today's technology, the method used to produce the largest quantities of hydrogen is the steam reformation of hydrocarbons, performed as a two-step process.   Because there are many different hydrocarbons, the empirical formula of (CH_2) will be used here for illustration, even though there are only a few specific hydrocarbons having that exact formula. The chemical reactions of the two-step process are as follows:

Reaction 1. $H_2O + (CH_2) \longrightarrow 2H_2 + CO$ conducted at high
                                                                temperatures
Reaction 2. $H_2O + CO \longrightarrow H_2 + CO_2$ conducted at low
                                                                temperatures

Reaction 1 is endothermic, and the gas must be continuously heated to drive the reaction. In the temperature range of 900° to 1000°C, reaction 1 is quite rapid and goes very nearly to completion. The gas produced by the reactor in which the reaction is carried out is cooled to 400° to 500°C, more steam is mixed with the gas, it is passed over a catalyst, and reaction 2 takes place. The simplified overall chemical reaction can be written as follows:

$$2H_2O + (CH_2) \longrightarrow 3H_2 + CO_2$$

From this reaction mixture, the hydrogen is separated and purified and supplied to the hydrogen-using process. The simplified chemistry is quite straightforward but, in actual practice, the process is more complex.[1]

The most desirable feedstock for hydrogen production is methane from natural gas, because of the higher hydrogen content of the methane molecule compared to other hydrocarbons. For the reaction using methane, the first reaction can be rewritten as follows:

$$H_2O + CH_4 \longrightarrow 3H_2 + CO$$

After the initial reaction to produce the mixture of hydrogen and carbon monoxide, reaction 2 is carried out as shown above. One of the main advantages of methane use is clear from examination of the overall reaction of steam with methane to form hydrogen:

$$2H_2O + CH_4 \longrightarrow 4H_2 + CO_2$$

The feedstock-to-hydrogen weight ratio is 2, with 5.5 kilograms of carbon dioxide produced per kilogram of hydrogen. By starting with methane, more total hydrogen is produced in the process stream than is produced with any other hydrocarbon. A second advantage of methane use accrues from the low level of sulfur compounds ordinarily found in natural gas and the ease of removing what sulfur is present. Thus, for the production of hydrogen from hydrocarbons, methane is the starting material of choice and is in fact the material most used. It is often possible, however, depending on the particular configuration of the plant, to substitute propane for methane, with some reduction in yield. This substitution is used on a temporary basis, but the future supply of propane is even more uncertain than the supply of methane, so propane will never replace methane as the major feedstock for hydrogen production.

The chemistry of hydrogen production from residual oils is essentially that given in reactions 1 and 2, but because the physical and, to some extent, the chemical properties of the residual oils are so much different than those of methane, the implementation of the process is rather different. For the residual oils, the stoichiometry of hydrogen to carbon gives an average formula that is more nearly $CH_{1.8}$. These oils are less reactive and less volatile than methane, which leads to the requirement that a portion of the oil be directly oxidized in the first stem to provide the higher temperature necessary for the first partial oxidation step. These considerations lead to a somewhat different series of reactions:

Reaction 3. $.65CH_{1.8} + .65H_2O \longrightarrow .65CO + 1.23H_2$

Reaction 4. $.35CH_{1.8} + .51O_2 \longrightarrow .35CO_2 + .32H_2O$

Reaction 5. $.65CO + .65H_2O \longrightarrow .65CO_2 + .65H_2$

for an overall reaction of

Reaction 6. $CH_{1.8} + .98H_2O + .51O_2 \longrightarrow CO_2 + 1.88H_2$

For this process, the feedstock-to-hydrogen ratio is 3.6, with 11.7 kilograms of carbon dioxide produced per kilogram of hydrogen. Only 5.5 percent as much hydrogen is produced with residual oils as feedstock as with methane, and 212 percent more carbon dioxide is produced. The lower reactivity of the residual oils in the reaction with water necessitates that the reaction be performed at a temperature of 1200° to $1300^\circ C$, $300^\circ C$ hotter than necessary for the production of hydrogen from methane. In summary, the need for greater quantities of feedstock, the higher temperatures required, and the increased emissions of carbon dioxide mitigate against the use of this process when sufficient natural gas is available.

COAL

Hydrogen is also manufactured from coal, and again the general overall process corresponds roughly to reactions 1 and 2. As with residual oils, the reaction requires the addition of oxygen to provide sufficient energy to drive the process. Coal has even less initial hydrogen than residual oils, and it does not have a uniform composition. There are at least three broad types of coal mined and used as fuels--anthracite, bituminous, and lignite--and among these three types are many variations and sub-categories. It is probably more informative to consider the coals a spectrum of materials ranging from coal with almost no volatile matter (anthracite) and no hydrogen to coal with a large amount of volatile matter and a molar hydrogen-to-carbon ratio approaching 1:1 (lignite). In addition to the variation in volatile components and hydrogen are large variations in the quantity of ash and its composition, the amount of sulfur and its form, and the physical properties of the coal as a solid. For this discussion, the amount of ash will be ignored because it has no direct effect on the production of hydrogen, and the composition will be assumed to be $CH_{0.8}$, a sort of grand average. The reaction of coal with water to produce hydrogen becomes

Reaction 7. $CH_{0.8} + .6H_2O + .7O_2 \longrightarrow CO_2 + H_2$

For this process, the feedstock-to-hydrogen ratio is 6.4, with 22 kilograms of carbon dioxide produced per kilogram of hydrogen--only 31 percent as much as produced with methane and 57 percent as much as with residual oils. The carbon dioxide production is 188 percent more than with residual oils and 400 percent more than with methane. In addition to this is the great disadvantage stemming from the solid form of coal and its ash. Solids are inherently difficult to handle in a chemical process because they cannot be pumped to high pressures; do not seal the entrance through which they are passed; and can pack, settle, and clump, making handling very difficult.[2,3]

Hydrogen has been produced from coal on a commercial scale in the past and is now being produced on a modest scale in South Africa. Production will undoubtedly begin within the United States if large-scale gasification and liquefaction of coal is undertaken. The most important processes currently in use are the Lurgi and Koppers-Totzek, illustrated in Figures 4-1 and 4-2. The primary difference between these processes is not the chemistry of the conversion of coal to hydrogen but the way in which the solid coal and ash are handled.

The need for movement of solids in and out of the reactor has several direct process consequences that make hydrogen production from coal less attractive than from other sources. The difficulty in sealing the solid inlet and outlet against high pressures results in the operation of all coal processes at relatively low pressures. This results in the need for very large reactors to process much gas and the use of extra energy to compress the hydrogen gas after it leaves the reactor so that it can be transferred to the next process or stored.

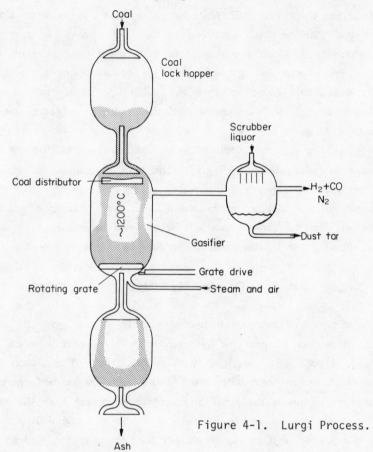

Figure 4-1. Lurgi Process.

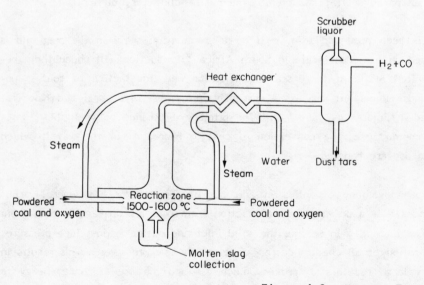

Figure 4-2. Koppers-Totzek Process.

A number of other variations on the coal gasification process are under investigation. They differ mainly in the manner in which the equipment is arranged and the materials handled. Two processes directed at the production of a gas mixture with a composition suitable for methane synthesis for use as natural gas are the Synthane process, under development by the U.S. Bureau of Mines, and the HYGAS process, under development by the Institute of Gas Technology.

The carbon dioxide acceptor process being developed by the Consolidated Coal Company is also directed at the production of a gas mixture suitable for the synthesis of methane. This process uses a somewhat different series of chemical reactions to accomplish the conversion of coal to a mixture of hydrogen and carbon monoxide. A natural mixture of calcium and magnesium carbonate (dolomite) is mixed with coal char, and air is forced through the mixture. The char combustion raises the temperature high enough so that the carbon dioxide is driven from the dolomite, leaving a mixture of calcium and magnesium oxides. While this mixture is still hot from the char combustion step, it is blended with more coal and steam is forced through the mixture. The hot dolomite supplies part of the energy necessary to cause the steam to react with the coal. The remaining energy is supplied by the reaction of the calcium and magnesium oxides with the product carbon dioxide to reform the carbonates. Sufficient excess coal is added in this step so that the expended carbonates contain enough residual coal char to drive the oxidation step starting this reaction. The advantage of this process lies in the separation of the water reduction step from the air oxidation step, which eliminates the contamination of the output gas stream by the nitrogen from the air.

The steam iron process is another process designed to separate the oxidation step from the hydrogen reduction step. In this process, under development by the Institute of Gas Technology, a mixture of air and steam is continuously forced through a bed of coal to produce a low-heat-value gas with hydrogen, carbon monoxide, and the nitrogen of the air. This gas mixture is used to reduce iron oxide to metallic iron. The hot metallic iron direct from the reduction step is treated with steam to produce iron oxide and hydrogen essentially uncontaminated with other gases. The iron oxide that is produced is recycled. The steam iron process has the great advantage of producing hydrogen of relatively high purity; it has the disadvantage of requiring the handling of a second solid, iron oxide.

These difficulties are the driving force behind the replacement of the earlier generation of coal gasification plants with processes using petroleum products. Unfortunately, the coming shortages of oil and natural gas are going to force us to again look at the production of hydrogen from coal.

ELECTROLYSIS

A significant amount of hydrogen is produced by the electrolysis of water, a simple electrochemical process that is well understood. In operation, it produces hydrogen and oxygen co-products of very high purity, and the only input required is pure water and electrical energy. Electrolysis is not used on a much wider basis because electrical energy is three to five times more expensive than the same quantity of energy derived directly from fossil fuels. Only in places where electrical energy is quite inexpensive or where supply makes fossil energy very expensive is there large-scale production of hydrogen by electrolysis. Some large electrolytic hydrogen production facilities are listed in Table 4-1. All are sited next to large hydroelectric facilities that produce low-cost electrical power.

Table 4-1. Electrolysis Plants

Plant Location	Capacity	
	Megawatts Electric	Cubic Meters/Hour Hydrogen
Rjukan, Norway	165	27,900
Glomfjord, Norway	160	27,100
Trail, Canada	90	15,200
Nangal, Egypt	125	21,100
Aswan, Egypt	100	16,900
Reykjavik, Iceland	20	3,000

Electrolysis is a very simple process. Figure 4-3 shows the critical parts of an electrolysis cell and identifies the sources of electrical resistance that affect its performance. In the ideal case, a voltage of 1.47 applied to a water electrolysis cell at 25°C would decompose the water into hydrogen and oxygen isothermally at 100-percent electrical efficiency. A voltage as low as 1.23 will decompose the water but, under these conditions, the reaction is endothermic and energy in the form of heat would be absorbed from the surroundings. At voltages higher than 1.47, the water is decomposed and heat is dissipated to the surroundings. To obtain maximum efficiency from the input energy, it is desirable to operate the electrolysis cell at a voltage only slightly above the minimum necessary for the formation of hydrogen and oxygen. Unfortunately, under these circumstances very little electric current flows, and the actual

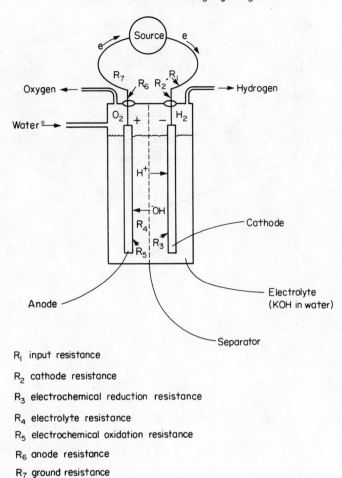

R$_1$ input resistance

R$_2$ cathode resistance

R$_3$ electrochemical reduction resistance

R$_4$ electrolyte resistance

R$_5$ electrochemical oxidation resistance

R$_6$ anode resistance

R$_7$ ground resistance

Figure 4-3. Electrolysis Cell.

production rate of the hydrogen measured per square centimeter of electrode surface is very low. If this technique were used for the production of hydrogen, the cells would have to be very large, and thus relatively costly, to produce a significant amount of hydrogen. As the voltage is increased, the production per unit area of cell increases, but the efficiency of use of the electric energy decreases. At higher temperatures, the voltage necessary for the decomposition of the water decreases so that, at a given voltage, the production rate per unit area increases. This effect is shown in Figure 4-4.

Ideally, it would be appropriate to operate the electrolysis cell at the highest possible temperature, but the vapor pressure of water makes operation more difficult at high temperatures. At 100°C, the vapor pressure of the water is, of course, equal to

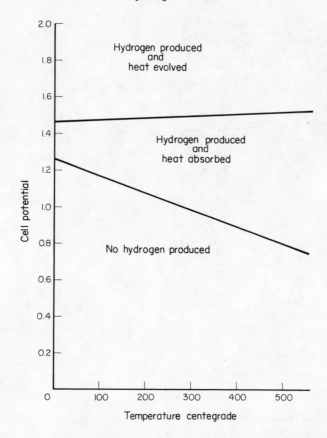

Figure 4-4.   Temperature-Voltage Relationship for Water
               Electrolysis.

atmospheric pressure.   Operation at temperatures higher than this requires that the cell
and all its support components be fabricated to operate at elevated pressures.
Elevating the temperature results in the reduction of the resistances represented in
Figure 4-3 as R_3 and R_5, resistances that result in the production of hydrogen and
oxygen by the performance of electrochemical work.   Reducing the other resistances to
the lowest possible value is also necessary for a high-efficiency cell; R_1, R_2, R_6, and
R_7 are all related to the electric circuit elements leading into the cell.   Although
these resistances are none the less important, they can become a minor part of the
losses in the cell by selecting materials that are good electrical conductors and making
the cell parts of an adequate size.

This leaves R_4 as the only other resistance under the control of the electrolysis
cell designer.   Pure water is not a very good conductor of electricity, as it has an
electrical resistance of over 10^6 ohms per centimeter.   A cell that uses only pure

water would pass so little current that almost no water would be electrolyzed. There are several options available for increasing the conductance of the water and, as a result, reducing resistance R_4. For example, the addition of any ionizable solute to the water will dramatically increase its ability to conduct current. There are literally hundreds of possible solutes that have the property of ionizing in water solutions. Virtually all salts, acids, and bases ionize to a greater or lesser extent in water and could serve to raise the current-carrying capacity of the water. Examples of this are as follows:

Salt:
$$NaCl \xrightarrow{\;H_2O\;} Na^+ + Cl^- \text{ in solution}$$

Acid:
$$H_2SO_4 \xrightarrow{\;H_2O\;} 2H^+ + SO_4^= \text{ in solution}$$

Base:
$$KOH \xrightarrow{\;H_2O\;} K^+ + OH^- \text{ in solution}$$

Water itself ionizes to a slight extent,

$$H_2O \longrightarrow H^+ + OH^-$$

but the product of the concentration of these two ions in very pure water is only 10^{-14}. This small amount of ionization is the reason that water conducts at all, but its low value results in the low value of conductance of water.

When a salt such as sodium chloride (common table salt) is added to the water, the resulting solution has a high conductivity because the mobile sodium and chlorine ions serve as current carriers. The use of salts does not work out in practice, how- ever, because there is no salt with high solubility having an anion and a cation that are stable to the electrochemical processes that are occurring at the electrolyte solu- tion interface, R_3 and R_5. With common salt, the tendency is for the chloride ions to be oxidized to produce hypochlorite ions:

$$Cl^- + 2OH^- \longrightarrow ClO^- + H_2O + 2e^-$$

The hypochlorite ions can decompose to produce either oxygen or chlorine and can further react to produce the higher oxides of chlorine. The chlorine produced can attack the metallic electrical-conducting electrodes, decreasing their life drastically.

The electrolysis of sodium chloride solutions is used for the production of chlorine industrially; there is, of course, a by-product production of hydrogen and sodium hydroxide. Typical of these processes are the Choralkali and the Solvay. The difficulty of obtaining long electrode life and the problem of disposal of the chlorine produced makes these processes unattractive for the large-scale production of hydrogen for use as a fuel.

The use of acids presents problems similar to those of salt use. As an example, with sulfuric acid there is a significant production of peroxymonosulfuric acid: $HOOSO_3H$. In addition, it is very difficult with acids to provide electrode materials that are adequate conductors of electricity and are also resistant to the corrosive effects of the strong acid solutions. Some success has been achieved with the design of acid cells using phosphoric acid as the electrolyte and platinum-based electrodes, but there is no significant commercial use of liquid electrolyte cells for the electrolysis of water using acid electrolytes.

Water solutions of chemical bases (primarily potassium hydroxide) are used in all current commercial electrolyzers. The simple electrolyzer configuration shown in Figure 4-3, using an asbestos separator and a metal electrolyte tank, represents the technology used in all the large electrolyzers listed in Table 4-1.

The alkaline electrolytes provide an excellent conductor between the anode and cathode and present minimum problems in use. The reaction at the cathode is as follows:

$$K^+ + e^- \longrightarrow K^o$$

$$K^o + H_2O \longrightarrow K^+ + \tfrac{1}{2}H_2 + OH^-$$

$$\tfrac{1}{2}H_2 + \tfrac{1}{2}H_2 \longrightarrow H_2$$

The positively charged potassium ion is the mobile charge carrier, and it migrates to the cathode because of its attraction to the negative charge on the cathode. The actual hydrogen ion concentration (H^+) is extremely low, less than 10^{-13} molar, so it is not possible to reduce the hydrogen ions directly. The potassium ion is present in concentrations of several moles per liter. Actual free potassium metal is not formed: the reduced potassium ion instantaneously reacts with the water of the solution to release a hydrogen atom and form a hydroxyl ion. The electron responsible for the cathode negative charge is transferred through the potassium ion to form one hydroxyl ion with a negative charge and a free hydrogen atom. Free hydrogen atoms are very reactive and bind tightly to the metal of the cathode. Two such atoms must come together and react to form a hydrogen molecule before they can escape as a gas. If an atomic layer of hydrogen atoms builds up on the surface of the electrode, the current flow is reduced and the rate of hydrogen production drops off or completely stops. On any electrode there will be a finite rate of combination of the hydrogen atoms to produce hydrogen gas, but on some metals the rate is much faster than on others. It is also possible to coat the surface of a metal that has a slow rate of hydrogen recombination with a catalyst to increase the rate of recombination and the rate of gaseous hydrogen production.

The best metals for the cathode are the metals in the platinum group. Unfortunately, these metals are very expensive, and their use as the electrode material would not be practical. Nickel, however, has been found to be almost as active as the platinum metals and is very much less costly. Nickel cathodes coated with very small quantities of platinum metals give very good recombination rates and are the most commonly used type of electrodes. A similar complex situation exists at the anode, where hydroxyl ions (OH^-) are attracted to the positive charge of the anode and react in the following manner:

$$OH^- \longrightarrow OH^0 + e^-$$

$$2OH^0 \longrightarrow H_2O + \tfrac{1}{2}O_2$$

$$\tfrac{1}{2}O_2 + \tfrac{1}{2}O_2 \longrightarrow O_2$$

As with the cathode, the reaction at the anode is complex and involves a reactive intermediate species that exists only for an exceedingly short period of time--the OH^0 species. Catalyzing its decomposition to O is not necessary, but again the combination of the two oxygen atoms to form an oxygen molecule is subject to improvement by catalysis. Nickel and copper both show activity in the recombination of the oxygen atoms, and anodes composed of these metals coated with trace layers of the oxides of metals such as manganese, tungsten, and ruthenium have shown good performance as anodes.[4]

Separator technology also requires special care. The separator must be stable in very strong solutions of potassium hydroxide and must be permeable to the flow of ions that are carrying the current across the cell without adding too much resistance. The separator must also keep the product hydrogen and oxygen from mixing, both to ensure the purity of the gas and to prevent the possible buildup of an explosive composition of hydrogen and oxygen. The explosive potential of hydrogen and oxygen mixtures is very great, and their prevention must be a major concern of the cell designer.

The separator cannot itself be an electrical conductor, for it would effectively become a new anode and cathode and the reactions would occur at the surface of the separator. The most commonly used separator is a tightly woven and pressed felt-like material made from asbestos. These felts are wetted by the potassium hydroxide electrolyte, which can then penetrate the pores in the felt to provide a path for the diffusion of the ions carrying the electrolysis current. The pores, however, are quite small, and the surface tension of the liquid across the pores that are in contact with the gas is sufficiently high that the gas cannot penetrate the felt at low pressure differentials.

The asbestos separators have performed well in the current technology electrolyzers, but they have several shortcomings that will limit their use in more advanced systems.[5] Asbestos is a mineral product showing a significant amount of batch-to-batch variation that is difficult to predict from examination of the raw material. The designer must thus design the electrolyzer conservatively enough to account for the poorest performing batch of asbestos or must perform life tests with separators fabricated from each new batch of material. The resistance of the asbestos to the potassium hydroxide solution decreases with temperature, so that it is not possible to push the operating temperature of electrolyzers using asbestos much above the current operating temperature of 60° to 80°C. Because the separation of the oxygen and hydrogen depends on the capillary forces operating in the pores of the felt, the differential pressure that the cell can tolerate is limited to 10 to 20 inches of water (0.03 to 0.06 atmosphere). This low pressure capability requires very careful control of the cell pressures, a level of control that becomes increasingly difficult to maintain as the operating pressure of the cells is increased.

Operation of electrolysis cells at high pressures is very desirable. At high pressures, the size of the gas bubbles is reduced at a fixed production rate. The smaller the bubbles, the less they block the electrode surface and the less they contribute to high resistance at R_4. When a cell is operated at high pressure, the energy necessary to compress the gas comes from the electrochemical process itself, such that a slightly higher voltage is necessary to perform electrolysis at high pressure than at low pressure. The energy for compression rides along with the energy for decomposing the water and operates at nearly 100-percent efficiency, unlike external compressors that are relatively much less efficient. The gas is produced at the pressure desired for the end use--a liquefier, pipeline, or the like. This not only eliminates the inefficient mechanical compression step but eliminates the cost of the compressors from the cost of the production plant.

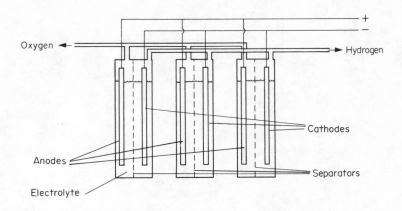

Figure 4-5.   Tank Type Electrolyzers

Individual cells can be packaged in two different ways in the hydrogen production plant: tank type or filter press type. In the tank-type electrolyzer (Figure 4-5), each cell is placed in its own container with a source of makeup water and low-voltage direct current electrical leads.   Because the cells are all in electrical parallel, they can be fed from one low-voltage bus bar, and the performance of one cell has little effect on the performance of the cell next to it.   If a cell should fail, it can simply be removed and repaired without significantly affecting the other cells or the produc-tion of the plant. The difficulty with this design arises from the requirement for very large current flows at low voltage.   Conductors for the high-current, low-voltage supply must be massive copper bus bars of very low resistance, or the losses occurring in R_1 and R_2 become excessive.   This problem is transmitted back to the power supply.   High-current, low-voltage power supplies require massive transformers and rectifiers. The requirements for these massive power supply components have mitigated against the use of the single cell, tank-type electrolyzers.   The tank-type electrolyzer also tends to require more floor space in the production plant than the filter press type.

The filter press type (Figure 4-6) is an electrical series type of electrolyzer, similar in appearance to the industrial filter press. It is constructed from a series of flat, plate-like cells stacked together like a pile of magazines.   The anode, the anode

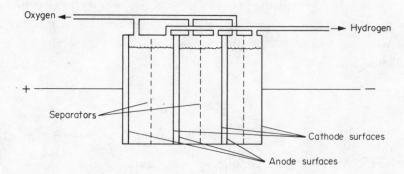

Figure 4-6. Filter Press Type Electrolyzers.

electrolyte space, the separator, the cathode electrolyte space, and the cathode make a sandwich. The back side of the cathode of one cell is the anode of the next cell. Each thin cell has its own water makeup supply and gas ports but no power supply attachment. Several of these cells are stacked together with the appropriate gas and water plumbing to make up a module. The voltage drop across the whole module is equal to the sum of the voltages for each cell. If, for example, 100 cells are stacked to produce the module, and each cell has a voltage requirement of 1.5 volts, the module itself can be supplied with a voltage of 100 x 1.5 = 150 volts. In theory, there is no limit to the number of cells that can be stacked together. The difficulty with this type of arrangement, however, is that each cell must be very nearly identical to the others, or it will be overloaded and fail. Should one cell fail, the whole module is out of service and must be disassembled to replace the single defective cell.

Both types of cell configuration have been used in the construction of large-scale electrolysis plants, but the trend is away from the tank type to the filter press type. The current major manufacturers of electrolysis units for plant operations are listed in Table 4-2.

Research in methods of improved electrolysis is quite active. Of major significance is a method using a solid polymer, Nafion®, which was developed by the Dupont Corporation. This polymer, a temperature-resistant solid, can replace the liquid electrolyte in the electrolysis cell. The polymer has strongly acid sulfonic acid groups attached to a polymer backbone that is fully fluorinated (similar to Teflon®). When the polymer is saturated with water, its acidity causes it to be an excellent conductor of electricity. Figure 4-7 shows the chemical structure of Nafion®. The charge carriers in this electrolyte are the hydrogen ions produced when the anode extracts hydroxyl ions from the water. These hydrogen ions, produced at the anode, travel

Table 4-2. Manufacturers of Electrolyzers

Lurgi Umwelt and Chemotechnik GmbH
6000 Frankfurt am Main
Gervinusstrasse, 17/19, Postfach 1191 81
Federal Republic of Germany

BBC Brown, Boveri & Co. Ltd.
CH 5401 Baden, Switzerland

Teledyne Energy Systems
110 West Timmonium Road
Timmonium, Maryland 21093
United States of America

Norsk Hydro a.a.
Bygdoy alle 2
Oslo 2, Norway

The Electrolyzer Corporation Limited
122 The West Mall
Etobicoke, Ontario, Canada M9C 1139

General Electric Company
Wilmington, Massachusetts 01887
United States of America

Oronzio De Nora
Impianti Elettrochmimici S.P.A.
Milan, Italy

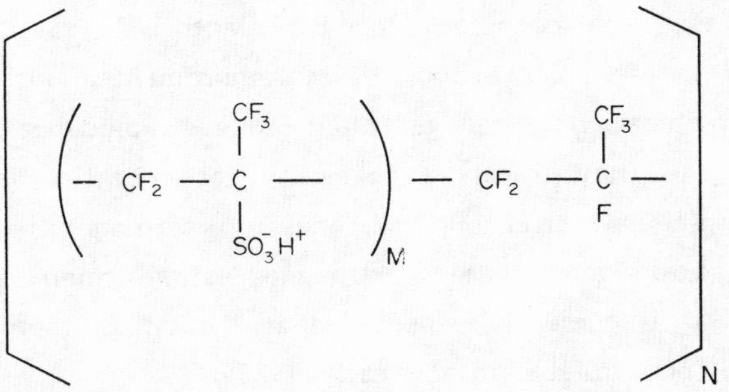

Figure 4-7. Chemical Structure of Nafion® Solid Polymer
            Electrolyte (Nafion® is a product of the
            E.I. Dupont Chemical Co.).

across the solid polymer electrolyte to the cathode, where they are discharged to produce molecular hydrogen. The hydroxyl ions produced at the anode are immediately
oxidized to produce oxygen gas.

There are several advantages to using the solid polymer electrolyte in the fabrication of electrolysis cells. Nafion® is a tough, dimensionally stable solid that can be
formed into thin sheets of uniform thickness, and it is impervious to gas. These properties allow it to be formed into electrolyte units of unusual thinness without any
separator. The minimum thickness combined with the high conductance for hydrogen
ions allows the value of R_4 in these cells to be much lower than it is in the traditional
type of cell. The high temperature stability coupled with the impermeability to gas
should make it possible to design cells that operate at high temperatures, above the
boiling point of water, and high pressures. The high temperature allows the cell to
operate further to the right on the voltage curve in Figure 4-4, thus improving the
electrical efficiency of the process. The high-pressure operation allows electrochemical
compression of the gas, reducing or eliminating the need for inefficient mechanical
compressors.

The General Electric Corporation, with the support of the U.S. Department of
Energy, has been performing research leading to the commercialization of electrolysis
cells using Nafion® as the electrolyte. The current plan is to have a unit capable of
producing hydrogen equivalent to 200 kilowatts of energy in the early 1980s. The
development challenges stem from the very properties that make Nafion® so desirable.
The solid nature of the electrolyte prevents the gas formed from forming bubbles, so
the electrode must be sufficiently porous that it can conduct the gas out of the cell.
This same porosity must allow for the entrance of the water vapor into the Nafion®.
In addition, the porous electrode must be forced into contact with the solid membrane
to provide the surface area necessary for the electrolysis to occur. The electrodes
must then withstand an often conflicting set of requirements: First, they must be good
activity for the recombination of the gas atoms to gas molecules; second, they must be
porous enough that the fluid transport necessary for operation is possible; and third,
they must possess sufficient structural toughness to withstand the high clamping pressures necessary to obtain good contact with the Nafion® membrane. These problems
are challenging, but it is currently expected that the difficulties will be overcome and
commercial cells will be available in the early 1980s.[6]

A number of corporations are actively manufacturing and marketing electrolyzer
systems. Figures 4-8 through 4-15 show the current state-of-the-art for the major
manufacturers. Of course, research is being performed to improve these electrolyzers,
which uniformly use water solutions of alkali as the electrolyte. The type of electro-

Figure 4-8.   BBC Brown, Boveri & Co. Ltd. Tank Type
              Electrolyzers During Assembly (Courtesy
              BBC Brown, Boveri & Co. Ltd., CH 5401 Baden,
              Switzerland).

Figure 4-9.   BBC Brown, Boveri & Co. Filter Press Type
Electrolyzers Installed for Use (Courtesy
BBC Brown, Boveri & Co. Ltd., CH 5401 Baden,
Switzerland).

Figure 4-10. Lurgi Filter Press Type Electrolyzers
Installed in a Hydrogen Production Facility
(Courtesy Lurgi Umwelt and Chemotechnik Gmblt,
600 Frankfurt am Main, Gervinusstrasse, 17/19,
Postfach 1191 81, Federal Republic of
Germany).

Figure 4-11. Teledyne HS Series Hydrogen Filter Press
Type Electrolysis Units for Producing Hydrogen
for Cooling Electric Generators and Semi-
conductor Manufacture (Courtesy Teledyne
Energy Systems, 110 West Timmonium Rd.,
Timmonium, Maryland 21093, U.S.A.).

Figure 4-12. Teledyne Energy Systems Electrolyzer Research
Test Bed, ARIES, Built for Brookhaven National
Laboratory Under Contract to the U.S.
Department of Energy (Courtesy Teledyne Energy
Systems, 110 West Timmonium Road, Timmonium,
Maryland 21093, U.S.A.).

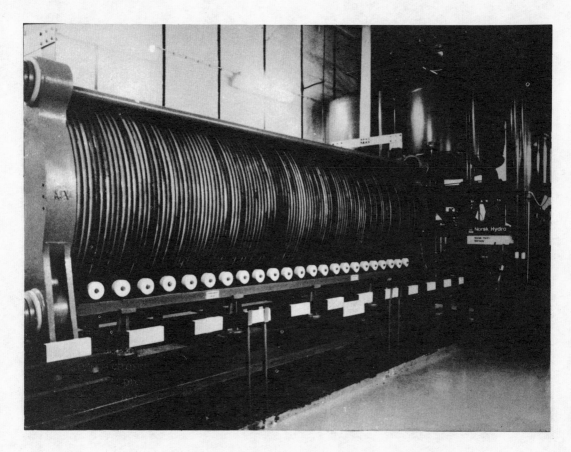

Figure 4-13. Norsk Hydro Electrolyzer (Filter Press Type)
in HCU-System. Side View, Showing the Cell
Block and Part of the Supporting Equipment.
Capacity Approximately 190 Nm³/hour (Courtesy
Norsk Hydro a.s Bygdoy alle 2, Oslo 2, Norway).

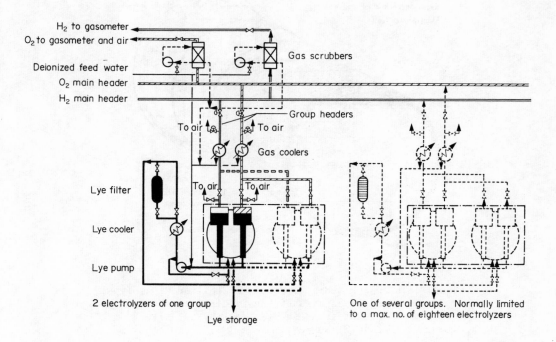

Basic flow diagram —— electrolysis plant in HCU system

Figure 4-14. Norsk Hydro Basic Flow Diagram Electrolysis
Plant in HCU System (Filter Press Type)
(Courtesy Norsk Hydro a.s Bygdoy alle 2,
Oslo 2, Norway).

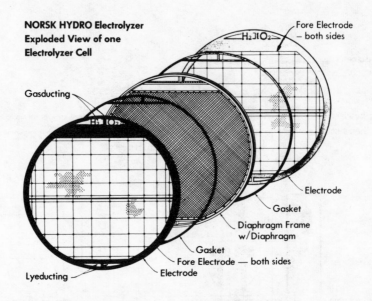

Figure 4-15. Norsk Hydro Electrolyzer (Filter Press Type)
Exploded View of One Electrolyzer Cell
(Courtesy Norsk Hydro a.s Bygdoy alle 2,
Oslo 2, Norway).

lyzer that will be used in the future will depend on the improvements in actual efficiency that result from current research.

THERMOCHEMICAL AND HYBRID PROCESSES

The lure of direct decomposition of water with thermal energy has intrigued many investigators, but the unavailability of stable construction materials capable of withstanding the temperature necessary to cause a significant portion of the water to thermally decompose (above 2500°C) has prevented any progress in this area. In the 1960s, a multistep system of thermally driven chemical reactions for the production of hydrogen was postulated by James E. Funk[7] of the University of Kentucky and Cesare Marchetti[8], currently at the International Institute for Applied Systems Analysis, Laxenburg, Austria. These reactions can be generalized as follows:

$$AB + H_2O + Heat \longrightarrow AH_2 + BO$$
$$AH_2 + Heat \longrightarrow A + H_2$$
$$2BO + Heat \longrightarrow 2B + O_2$$
$$A + B + Heat \longrightarrow AB$$

It was hoped that a series of reactions would be found that could be driven by temperatures within the range of those commonly used by industry. This would in effect sidestep the very high temperature necessary for the direct thermal splitting of water and provide a technique for the production of hydrogen and oxygen from water by a purely thermal chemical method.

It was recognized from the onset that there were a number of very stringent conditions that would have to be applied to these reaction series if they were to be successful; they are as follows:

- The yield of each reaction in the sequence must be quite high. The overall yield of the sequence would be the product of the yields of each step. If there were four steps, as shown in the idealized reaction sequence, each with a yield of 90 percent, the overall yield would be

$$.9 \times .9 \times .9 \times .9 = .65 \text{ or } 65 \text{ percent}$$

If the individual reactions had yields of only 80 percent, the overall yield would be

$$.8 \times .8 \times .8 \times .8 = .41 \text{ or } 41 \text{ percent}$$

Any reactions with low yields would so strongly affect the overall yield of the process that it would be very difficult to accommodate them.

- The number of individual reactions must be a minimum. The more reactions in the sequence, the more important the yield of each reaction becomes. For a sequence of 10 steps, a 90-percent yield per step would give an overall yield of only 34 percent, and an 80-percent yield would give an overall yield of only 11 percent.

- The intermediate products that required high temperature processing would have to be relatively easy to handle. If intermediate materials were difficult to handle, the cost of the process could become prohibitive.

- The reactions could not result in chemical by-products. The production of a by-product in one of the reactions would necessitate the addition of a separate process sequence to recycle the material diverted to the formation of the by-products. This separate sequence would be effectively the same as the addition of another reaction step to the overall process.

• The components A and B must be elements or compounds that are available in reasonable quantities at reasonable cost.

• None of the compounds involved in the process can present an unreasonable hazard to the environment.

Since the mid-1960s, research has been performed to investigate a number of potential thermochemical cycles for the production of hydrogen. These research efforts have ranged from pure paper studies of potential reaction sequences, to the investigation of specific chemical reactions, to the fabrication of laboratory bench-scale pilot plants for investigating the total process. Table 4-3 lists the major reaction sequences that have been investigated beyond the stage of theoretical investigations. Of these 54 reaction sequences, only a few are still thought to have any possible chance for future industrial use[9].

The General Atomic Company of San Diego, California, is planning a laboratory-scale demonstration plant that should be operating in the early 1980s. Their process uses the following set of reactions:

$$2H_2O + SO_2 + xI_2 \longrightarrow H_2SO_4 + 2HI_x \text{ (in water solution)}$$

$$2HI_x \longrightarrow xI_2 + H_2 \quad (300^{\circ}C)$$

$$H_2SO_4 \longrightarrow H_2O + SO_2 + \tfrac{1}{2}O_2 \; (870^{\circ}C)$$

This reaction sequence meets most of the criteria established for a successful thermo-chemical hydrogen production reaction, but there are still many problems and significant doubt whether the overall process will be successful.[10]

The separation of the water solution of the two acids produced in the first reaction is a problem area. At fairly high concentrations of acid, the sulfuric acid and the hydropolyiodic acid are not mutually soluble and separate into an upper and lower layer, with the lower layer being much richer in the hydropolyiodic acid. This behavior allows the gross separation of the two acids. Unfortunately, the phase separation does not give perfect separation, and a little sulfuric acid remains in the hydropolyiodic acid and vice versa. Improvements in the separation of the acid are under investigation, as are methods of handling the presence of small amounts of one acid as an impurity in the other, further along in the process stream.

Table 4-3. Summary of the Most Significant Thermochemical Water Decomposition Cycles

Process	Reaction	Process	Reaction
Mark 1 (Italian)	1) $CaBr_2 + 2H_2O \xrightarrow{730°C^*} Ca(OH)_2 + 2HBr$	Mark 2 1972 (Italian)	1) $Mn_2O_3 + 4NaOH \xrightarrow{800°} 2Na_2O \cdot MnO_2 + H_2O + H_2$
	2) $2HBr + Hg \xrightarrow{250°} HgBr_2 + H_2$		2) $2Na_2O \cdot MnO_2 + nH_2O \xrightarrow{100°} 4NaOH(aq) + 2MnO_2$
	3) $HgBr_2 + Ca(OH)_2 \xrightarrow{200°} CaBr_2 + HgO + H_2O$		3) $2MnO_2 \xrightarrow{600°} Mn_2O_3 + \frac{1}{2}O_2$
	4) $HgO \xrightarrow{600°} Hg + \frac{1}{2}O_2$		
Mark 1B (Italian)	1) $CaBr_2 + 2H_2O \xrightarrow{730°} Ca(OH)_2 + 2HBr$	Mark 2C (Italian)	1) $Mn_2O_3 + 2Na_2CO_3 \xrightarrow{850°} 2Na_2O \cdot MnO_2 + CO_2 + CO$
	2) $2HBr + Hg_2Br_2 \xrightarrow{120°} 2HgBr_2 + H_2$		2) $CO + H_2O \xrightarrow{500°} H_2 + CO_2$
	3) $HgBr_2 + Hg \xrightarrow{120°} Hg_2Br_2$		3) $2Na_2O \cdot MnO_2 + nH_2O + 2CO_2 \xrightarrow{100°} 2Na_2CO_3(aq) + 2MnO_2$
	4) $HgBr_2 + Ca(OH)_2 \xrightarrow{200°} CaBr_2 + HgO + H_2O$		4) $2MnO_2 \xrightarrow{600°} Mn_2O_3 + \frac{1}{2}O_2$
	5) $HgO \xrightarrow{600°} Hg + \frac{1}{2}O_2$		
Mark 1C (Italian)	1) $2CaBr_2 + 4H_2O \xrightarrow{730°} 2Ca(OH)_2 + 4HBr$	Mark 3 (Italian)	1) $Cl_2 + H_2O \xrightarrow{800°} 2HCl + \frac{1}{2}O_2$
	2) $4HBr + Cu_2O \xrightarrow{100°} 2CuBr_2 + H_2O + H_2$		2) $2HCl + 2VOCl \xrightarrow{170°} 2VOCl_2 + H_2$
	3) $2CuBr_2 + Ca(OH)_2 \xrightarrow{100°} 2CuO + 2CaBr_2 + 2H_2O$		3) $4VOCl_2 \xrightarrow{600°} 2VOCl + 2VOCl_3$
	4) $2CuO \xrightarrow{900°} Cu_2O + \frac{1}{2}O_2$		4) $2VOCl_3 \xrightarrow{200°} 2VOCl_2 + Cl_2$
Mark 1S (Italian)	1) $SrBr_2 + H_2O \xrightarrow{800°} SrO + 2HBr$	Mark 4 (Italian)	1) $Cl_2 + H_2O \xrightarrow{800°} 2HCl + \frac{1}{2}O_2$
	2) $2HBr + Hg \xrightarrow{200°} HgBr_2 + H_2$		2) $2HCl + S + 2FeCl_2 \xrightarrow{100°} H_2S + 2FeCl_3$
	3) $SrO + HgBr_2 \xrightarrow{500°} SrBr_2 + Hg + \frac{1}{2}O_2$		3) $H_2S \xrightarrow{800°} H_2 + \frac{1}{2}S_2$
			4) $2FeCl_3 \xrightarrow{420°} 2FeCl_2 + Cl_2$

Table 4-3. Summary of the Most Significant Thermochemical
Water Decomposition Cycles (Continued)

Process	Reaction	Process	Reaction
Mark 5	1) $CaBr_2 + H_2O + CO_2 \xrightarrow{600°} CaCO_3 + 2HBr$	Mark 7A	1) $6FeCl_2 + 8H_2O \xrightarrow{650°} 2Fe_3O_4 + 12HCl + 2H_2$
	2) $CaCO_3 \xrightarrow{900°} CaO + CO_2$		2) $2Fe_3O_4 + \frac{1}{4}O_2 \xrightarrow{350°} 3Fe_2O_3$
	3) $2HBr + Hg \xrightarrow{250°} HgBr_2 + H_2$		3) $2Fe_2O_3 + 12HCl \xrightarrow{120°} 4FeCl_3 + 6H_2O$
	4) $HgBr_2 + CaO + nH_2O \xrightarrow{200°} CaBr_2(aq) + HgO$		4) $Fe_2O_3 + 3Cl_2 \xrightarrow{1000°} 2FeCl_3 + \frac{3}{2}O_2$
(Italian)	5) $HgO \longrightarrow Hg + \frac{1}{2}O_2$	(Italian)	5) $6FeCl_3 \xrightarrow{420°} 6FeCl_2 + 3Cl_2$
Mark 6	1) $Cl_2 + H_2O \xrightarrow{800°} 2HCl + \frac{1}{2}O_2$	Mark 7B	1) $6FeCl_2 + 8H_2O \xrightarrow{650°} 2Fe_3O_4 + 12HCl + 2H_2$
	2) $2HCl + 2CrCl_2 \xrightarrow{170°} 2CrCl_3 + H_2$		2) $2Fe_3O_4 + \frac{1}{4}O_2 \xrightarrow{350°} 3Fe_2O_3$
	3) $2CrCl_3 + 2FeCl_2 \xrightarrow{700°} 2CrCl_2 + 2FeCl_3$		3) $3Fe_2O_3 + 9Cl_2 \xrightarrow{1000°} 6FeCl_3 + \frac{3}{2}O_2$
(Italian)	4) $2FeCl_3 \xrightarrow{350°} 2FeCl_2 + Cl_2$		4) $6FeCl_3 \xrightarrow{420°} 6FeCl_2 + 3Cl_2$
		(Italian)	5) $12HCl + 3O_2 \xrightarrow{400°} 6Cl_2 + 6H_2O$
Mark 6C	1) $Cl_2 + H_2O \xrightarrow{800°} 2HCl + \frac{1}{2}O_2$	Mark 8	1) $6MnCl_2 + 8H_2O \xrightarrow{700°} 2Mn_3O_4 + 12HCl + 2H_2$
	2) $2HCl + 2CrCl_2 \xrightarrow{170°} 2CrCl_3 + H_2$		2) $3Mn_3O_4 + 12HCl \xrightarrow{100°} 6MnCl_3 + 3MnO_2 + 6H_3O$
	3) $2CrCl_3 + 2FeCl_2 \xrightarrow{700°} 2CrCl_2 + 2FeCl_3$	(Italian)	3) $3MnO_2 \xrightarrow{900°} Mn_3O_4 + O_2$
	4) $2FeCl_3 + 2CuCl \xrightarrow{150°} 2FeCl_2 + 2CuCl_2$		
(Italian)	5) $2CuCl_2 \longrightarrow 2CuCl + Cl_2$		
Mark 7	1) $6FeCl_2 + 8H_2O \xrightarrow{650°} 2Fe_3O_4 + 12HCl + 2H_2$	Mark 9	1) $6FeCl_2 + 8H_2O \xrightarrow{450°} 2Fe_3O_4 + 12HCl + 2H_2$
	2) $2Fe_3O_4 + \frac{1}{4}O_2 \xrightarrow{350°} 3Fe_2O_3$		2) $2Fe_3O_4 + 3Cl_2 + 12HCl \xrightarrow{150°} 6FeCl_3 + 6H_2O + O_2$
	3) $3Fe_2O_3 + 18HCl \xrightarrow{120°} 6FeCl_3 + 9H_2O$	(Italian)	3) $6FeCl_3 \xrightarrow{420°} 6FeCl_2 + 3Cl_2$
	4) $6FeCl_3 \xrightarrow{420°} 6FeCl_2 + 3Cl_2$		
(Italian)	5) $3Cl_2 + 3H_2O \xrightarrow{800°} 6HCl + \frac{3}{2}O_2$		

Table 4-3. Summary of the Most Significant Thermochemical
Water Decomposition Cycles (Continued)

Process	Reaction	Process	Reaction
Argonne (US)	1) $LiNO_2 + I_2 + H_2O \xrightarrow{27°} LiNO_3 + 2HI$	Agnes (US)	1) $3FeCl_2 + 4H_2O \xrightarrow{450°\sim750°} F_3O_4 + 6HCl + H_2$
	2) $2HI \xrightarrow{427°} I_2 + H_2$		2) $Fe_3O_4 + 8HCl \xrightarrow{100°\sim110°} FeCl_2 + 2FeCl_3 + 4H_2O$
	3) $LiNO_3 \xrightarrow{427°} LiNO_2 + \frac{1}{2}O_2$		3) $2FeCl_3 \xrightarrow{300°} 2FeCl_2 + Cl_2$
			4) $Cl_2 + Mg(OH)_2 \xrightarrow{50°\sim90°} MgCl_2 + \frac{1}{2}O_2 + H_2O$
			5) $MgCl_2 + 2H_2O \longrightarrow Mg(OH)_2 + 2HCl$
Aerojet-General (US)	1) $2H_2O + 2Cs \xrightarrow{450°} 2CsOH + H_2$	Catherine (US)	1) $3I_2 + 6LiOH \xrightarrow{100°\sim190°} 5LiI + LiIO_3 + 3H_2O$
	2) $2CsOH \xrightarrow{250°} 2CsO_2 + H_2O$		2) $LiIO_3 + KI \xrightarrow{0} KIO_3 + LiI$
	3) $2CsO_2 \xrightarrow{450°} Cs_2O + \frac{3}{2}O_2$		3) $KIO_3 \xrightarrow{650°} KI + 1\frac{1}{2}O_2$
	4) $Cs_2O \longrightarrow 2Cs + \frac{1}{2}O_2$		4) $6LiI + 6H_2O \xrightarrow{450°\sim600°} 6HI + 6LiOH$
			5) $6HI + 3Ni \xrightarrow{150°} 3NiI_2 + 3H_2$
			6) $3NiI_2 \xrightarrow{700°} 3Ni + 3I_2$
GE	1) $2Cu + 2HCl \xrightarrow{100°} 2CuCl + H_2$	IGT C-5 (US)	1) $Fe_3O + 3SO_4 + H_2O \xrightarrow{500°} 3FeSO_2 + 2H_2$
	2) $4CuCl \xrightarrow{30°\sim100°} 2CuCl_2 + 2Cu$		2) $3FeSO_4 \xrightarrow{1000°} \frac{3}{2}Fe_2O_3 + \frac{3}{2}SO_2 + \frac{3}{2}SO_3$
Beulah (US)	3) $2CuCl_2 \xrightarrow{500°\sim600°} 2CuCl + Cl_2$		3) $\frac{3}{2}Fe_2O_3 + \frac{1}{2}SO_2 \xrightarrow{350°} Fe_3O_4 + \frac{1}{2}SO_3$
	4) $Cl_2 + Mg(OH)_2 \xrightarrow{80°} MgCl_2 + H_2O + \frac{1}{2}O_2$		4) $2SO_3 \xrightarrow{700°} 2SO_2 + O_2$
	5) $MgCl_2 + 2H_2O \xrightarrow{350°} Mg(OH)_2 + 2HCl$		
Jülich Center	1) $3FeO + H_2O \xrightarrow{200°} Fe_3O_4 + H_2$	IGT A-2 (US)	1) $3Fe + 4H_2O \xrightarrow{500°} Fe_3O_4 + 4H_2$
			2) $Fe_3O_4 + \frac{9}{2}Cl_2 \xrightarrow{1000°} 3FeCl_3 + 2O_2$
EOS (West German)	2) $Fe_3O_4 + FeSO_4 \xrightarrow{800°} 3Fe_2O_3 + 3SO_2 + \frac{1}{2}O_2$		3) $FeCl_3 \xrightarrow{350°} FeCl_2 + \frac{1}{2}Cl_2$
			4) $2FeCl_2 + 3H_2 \xrightarrow{1000°} 3Fe + 6HCl$
	3) $3Fe_2O_3 + 3SO_2 \xrightarrow{200°} 3FeSO_4 + 3FeO$		5) $6HCl \xrightarrow{500°} 3Cl_2 + 3H_2O$

Table 4-3. Summary of the Most Significant Thermochemical
Water Decomposition Cycles (Continued)

Process	Reaction	Process	Reaction
GE (US)	1) $2VCl_2 + 2HCl \xrightarrow{25°} 2VCl_3 + H_2$	Lawrence Livermore Lab. Univ. of Cali. (US)	1) $MgSe + 2H_2O \xrightarrow{100°} Mg(OH)_2 + H_2Se$
	2) $4VCl_3 \xrightarrow{700°} 2VCl_4 + 2VCl_2$		2) $2H_2Se \xrightarrow{200°} 2H_2 + 2Se$
	3) $2VCl_4 \xrightarrow{25°} 2VCl_3 + Cl_2$		3) $2Se + Mg(OH)_2 \xrightarrow{225°} H_2Se + \tfrac{1}{2}MgSe + \tfrac{1}{2}MgSeO_4$
	4) $H_2O + Cl_2 \xrightarrow{700°} 2HCl + \tfrac{1}{2}O_2$		$\tfrac{1}{2}MgSe + 2H_2O \longrightarrow 2H_2 + \tfrac{1}{2}MgSeO_4$
Aachen Univ. Jülich 1972 (West German)	1) $2CrCl_2 + 2HCl \xrightarrow{200°} 2CrCl_3 + H_2$	Lawrence Livermore Lab. Univ. of Cali. (US)	1) $2CsOH + \frac{x+1}{2}O_2 \xrightarrow{410°} 2CsO_x + H_2O$
	2) $2CrCl_3 \xrightarrow{800°} 2CrCl_2 + Cl_2$		2) $CsO_x + (x+y)Hg \xrightarrow{300°} CsHgy + xHgO$
	3) $H_2O + Cl_2 \xrightarrow{700°} 2HCl + \tfrac{1}{2}O_2$		3) $HgO \longrightarrow Hg + \tfrac{1}{2}O_2$
			4) $CsHgy + H_2O \longrightarrow Hg + CsOH + \tfrac{1}{2}H_2$
Euratom 1970	1) $C + H_2O \xrightarrow{700°} CO + H_2$	Lawrence Livermore Lab. Univ. of Cali. (US)	1) $CH_4 + H_2O \xrightarrow{700°} CO + 3H_2$
De Beni (Italian)	2) $CO + 2Fe_3O_4 \xrightarrow{250°} C + 3Fe_2O_3$		2) $CO + 2H_2 \xrightarrow{230°} CH_3OH$
	3) $3Fe_2O_3 \xrightarrow{1400°} 2Fe_3O_4 + \tfrac{1}{2}O_2$		3) $CH_3OH + As_2O_4 \xrightarrow{227°} CH_4 + As_2O_5$
			4) $\tfrac{1}{2}As_2O_5 \xrightarrow{700°} \tfrac{1}{2}As_2O_3 + \tfrac{1}{2}O_2$
			5) $\tfrac{1}{2}As_2O_3 + \tfrac{1}{2}As_2O_3 \xrightarrow{\;?\;} As_2O_4$
Lawrence Livermore Lab. Univ. of Cali. (US)	1) $K_2Se + 2H_2O \xrightarrow{100°} 2KOH + H_2Se$	General Atomic Company San Diego (US)	1) $2EuO + H_2O \xrightarrow{390°} H_2 + Eu_2O_3$
	2) $H_2Se \xrightarrow{200°} H_2 + Se$		2) $I_2 + SrO \xrightarrow{323°} \tfrac{1}{2}O_2 + SrI_2$
	3) $\tfrac{3}{2}Se + 2KOH \xrightarrow{700°} K_2Se + \tfrac{1}{2}SeO_2 + H_2O$		3) $Eu_2O_3 + SrI_2 \longrightarrow 2EuO + I_2 + SrO$
	4) $V_2O_4 + \tfrac{1}{2}SeO_2 \xrightarrow{327°} V_2O_5 + \tfrac{1}{2}Se$		
	5) $V_2O_5 \xrightarrow{500°} V_2O_4 + \tfrac{1}{2}O_2$		

Table 4-3. Summary of the Most Significant Thermochemical Water Decomposition Cycles (Continued)

Process	Reaction	Process	Reaction
(US)	1) $H_2O + Cl_2 \xrightarrow{700°} 2HCl + \frac{1}{2}O_2$	Argonne Nat. Lab. (US)	1) $H_2O + NH_3 + CO_2 + NaBr \xrightarrow{27°} NaHCO_3 + NH_4Br$
	2) $HCl + 2CuCl \xrightarrow{200°} 2CuCl_2 + H_2$		2) $NaHCO_3 \xrightarrow{127°} \frac{1}{2}Na_2CO_3 + \frac{1}{2}H_2O$
	3) $2CuCl_2 \longrightarrow 2CuCl + Cl_2$		3) $NH_4Br + Ag \xrightarrow{477°} \frac{1}{2}H_2 + NH_3 + AgBr$
			4) $AgBr + \frac{1}{2}Na_2CO_3 \xrightarrow{727°} Ag + NaBr + \frac{1}{2}CO_2 + \frac{1}{2}O_2$
IGT 1969 (US)	1) $Fe + H_2O \longrightarrow FeO + H_2$	THEME S-3 39 (US)	1) $SO_2 + H_2O + I_2 \longrightarrow SO_3 + 2HI$
	2) $3FeO + H_2O \xrightarrow{550°} Fe_3O_4 + H_2$		2) $SO_3 \longrightarrow SO_2 + \frac{1}{2}O_2$
	3) $Fe_3O_4 + CO \xrightarrow{950°} 3FeO + CO_2$		3) $2HI \longrightarrow H_2 + I_2$
	4) $FeO + CO \longrightarrow Fe + CO_2$		
	5) $2CO_2 \xrightarrow{315°} 2CO + O_2$		
Euratom 1972 (Italian)	1) $H_2O + Cl_2 \xrightarrow{700°} 2HCl + \frac{1}{2}O_2$	Argonne (US)	1) $2KNO_3 + I_2 \longrightarrow 2KI + 2NO_2 + O_2$
	2) $2HCl + 2FeCl_2 \xrightarrow{600°} 2FeCl_3 + H_2$		2) $2NO_2 + \frac{1}{2}O_2 + H_2O \longrightarrow 2HNO_3$
	3) $FeCl_3 \xrightarrow{350°} FeCl_2 + Cl_2$		3) $2HNO_3 + 2NH_3 \longrightarrow 2NH_4NO_3$
			4) $2KI + 2NH_4NO_3 \longrightarrow 2KNO_3 + 2NH_4I$
			5) $2NH_4I \longrightarrow 2NH_3 + I_2 + H_2$
Sourian Gaz de France 1972 (French)	1) $Sn + 2H_2O \xrightarrow{400°} 2H_2 + SnO_2$	West Germany	1) $6CO + 6H_2O \xrightarrow{3°\sim450°} 6CO_2 + 6H_2$
	2) $2SnO_2 \xrightarrow{1700°} 2SnO + O_2$		2) $6CO_2 + 6SO_2 + 6H_2O \xrightarrow{3°\sim350°} 6CO + 6H_2SO_4$
	3) $2SnO \xrightarrow{700°} SnO_2 + Sn$		3) $6H_2SO_4 + 2Fe_2O_3 \xrightarrow{\sim150 \text{ atm}} 2Fe_2(SO_4)_3 + 6H_2O$
			4) $2Fe_2(SO_4)_3 \xrightarrow{5°\sim600°} 2Fe_2O_3 + 6SO_3$
			5) $6SO_3 \xrightarrow{800°} 6SO_2 + 3O_2$

Table 4-3. Summary of the Most Significant Thermochemical
Water Decomposition Cycles (Continued)

Process	Reaction	Process	Reaction
Funk (US)	1) $H_2O + Cl_2 \longrightarrow 2HCl + \frac{1}{2}O_2$	Gaz de France (French)	1) $2KOH + 2K \xrightarrow{725^\circ} 2K_2O + H_2$
	2) $2TaCl_2 + 2HCl \longrightarrow 2TaCl_3 + H_2$		2) $K_2O_2 + H_2O \xrightarrow{25^\circ \sim 125^\circ} 2KOH + \frac{1}{2}O_2$
	3) $2TaCl_3 \longrightarrow 2TaCl_2 + Cl_2$		3) $2K_2O \xrightarrow{825^\circ} K_2O_2 + 2K$
LASL (US)	1) $SO_2 + 2H_2O + Br_2 \xrightarrow{100^\circ} H_2SO_4 + 2HBr$	Yokohama Mark 3 1973 (Japanese)	1) $2FeSO_4 + I_2 + 2H_2O \longrightarrow 2Fe(OH)SO_4 + 2HI$
	2) $H_2SO_4 \xrightarrow{900^\circ} H_2O + SO_2 + \frac{1}{2}O_2$		2) $2HI \longrightarrow H_2 + I_2$
	3) $2HBr + 2CrBr_2 \longrightarrow 2CrBr_3 + H_2$		3) $2Fe(OH)SO_4 \xrightarrow{250^\circ} 2FeSO_4 + H_2O + \frac{1}{2}O_2$
	4) $2CrBr_3 \longrightarrow 2CrBr_2 + Br_2$		
LASL (US)	1) $U_3O_8 + H_2O + 3CO_2 \longrightarrow 3UO_2CO_3 + H_2$	Hallett Air Products 1965 (US)	1) $H_2O + Cl_2 \xrightarrow{700^\circ} 2HCl + \frac{1}{2}O_2$
	2) $3UO_2CO_3 \longrightarrow 3UO_3 + 3CO_2$		2) $2HCl \xrightarrow{300^\circ} H_2 + Cl_2 (Electrolysis)$
	3) $3UO_3 \longrightarrow U_3O_8 + \frac{1}{2}O_2$		
LASL (US)	1) $6LiOH + 2MN_3O_4 \xrightarrow{5^\circ \sim 700^\circ} 3Li_2O \cdot Mn_2O_3 + 2H_2O + H_2$	LASL (US)	1) $H_2O + Cl_2 \xrightarrow{700^\circ} 2HCl + \frac{1}{2}O_2$
	2) $3Li_2O \cdot Mn_2O_3 + 3H_2O \xrightarrow{25^\circ \sim 80^\circ} 6LiOH + 3Mn_2O_3$		2) $2Hg + 2HCl \xrightarrow{300^\circ} 2HgCl + H_2$
	3) $3Mn_2O_3 \xrightarrow{800^\circ \sim 1000^\circ} 2Mn_3O_4 + \frac{1}{2}O_2$		3) $2HgCl \xrightarrow{500^\circ} 2Hg + Cl_2 (Electrolysis)$

Table 4-3. Summary of the Most Significant Thermochemical
Water Decomposition Cycles (Continued)

Process	Reaction	Process	Reaction
Hallett Air Products 1965 (US)	1) $H_2O + Cl_2 \xrightarrow{700°} 2HCl + \tfrac{1}{2}O_2$ 2) $2HCl + FeCl_2 \xrightarrow{200°} 2FeCl_3 + H_2(Electrolysis)$ 3) $2FeCl_3 + 2NO \xrightarrow{176°} 2FeCl_2 + NOCl$ 4) $2NOCl \xrightarrow{150°} 2NO + Cl_2$	Shell Process (US)	1) $6Cu + 3H_2O \xrightarrow{500°} 3Cu_2O + 3H_2$ 2) $Cu_2S + SO_2 + 3O_2 \xrightarrow{300°} 2CuSO_4$ 3) $2Cu_2S + 2CuSO_4 \xrightarrow{500°} 6Cu + 4SO_2$
(US)	1) $H_2O + Cl_2 \xrightarrow{700°} 2HCl + \tfrac{1}{2}O_2$ 2) $2HCl + 2CuCl \xrightarrow{200°} 2CuCl_2 + H_2(Electrolysis)$ 3) $2CuCl_2 \xrightarrow{300°} 2CuCl + Cl_2$	Tokyo Inst. of Technology 1975 (Japanese)	1) $6CaO + 6I_2 \xrightarrow{20°\sim30°} Ca(IO_3)_2 + 5CaI_2$ 2) $Ca(IO_3)_2 \xrightarrow{550°\sim700°} CaO + I_2 + \tfrac{5}{2}O_2$ 3) $5CaI_2 + 5H_2O \longrightarrow 5CaO + 10HI$ 4) $10HI \longrightarrow 5H_2 + 5I_2$
Ispra (Italian)	1) $H_2SO_3 + H_2O \longrightarrow H_2SO_4 + H_2(Electrolysis)$ 2) $H_2SO_4 \xrightarrow{800°} SO_3 + H_2O$ 3) $SO_3 \xrightarrow{800°} SO_2 + \tfrac{1}{2}O_2$ 4) $SO_2 + H_2O \xrightarrow{25°} H_2SO_3$	Osaka Inst. of Technology 1975 (Japanese)	1) $2NH_4I \longrightarrow 2NH_3 + H_2 + I_2$ 2) $BaCO_3 + I_2 \longrightarrow BaI_2 + CO_2 + \tfrac{1}{2}O_2$ 3) $BaI_2 + 2NH_3 + CO_2 + H_2O \longrightarrow BaCO_3 + 2NH_4I$

Hydrogen polyiodide is an unstable substance that decomposes to hydroiodic acid and free iodine, slightly at room temperature and more rapidly at higher temperatures:

$$HI_x \longrightarrow HI + \tfrac{1}{2}(x-1)I_2$$

This characteristic requires that the container in which the material is handled and processed must not only be resistant to acids (hydroiodic acid is a highly ionized and strong acid), but must also be resistant to the strongly oxidative and corrosive free iodine. The only substance identified to date that gives satisfactory service is fused quartz, an impractical substance for the fabrication of a production plant.

Finding materials capable of containing the decomposition of the sulfuric acid reaction is also a problem. Sulfuric acid is a strong acid, quite corrosive to many materials, and is made more so by the presence of residual hydropolyiodic acid and free iodine. As with the decomposition of hydroiodic acid, quartz is a suitable substance, but there are no other obvious candidates. In the sulfuric acid decomposition section, operating at 870°C, the materials must not only resist the acid attack but must be resistant to attack from the pure oxygen produced by the decomposition of the sulfuric acid.

The yields of these reactions are quite high, and there are no significant side reactions. The potential efficiency of the reaction is therefore high enough to be of interest for a hydrogen production method. Iodine, however, is not a plentiful element, and its cost is high. Traces of iodine (in the low part-per-million range) will probably be present in the hydrogen produced, which could render the hydrogen unsuitable for many applications and could result in severe corrosion of the equipment used to handle the hydrogen. The oxygen produced could contain traces of sulfur dioxide that would be difficult to remove completely. This trace contaminant could make the oxygen less than desirable for use and could even interfere with its disposal to the atmosphere. It is not clear at the present time if the problems surrounding the use of this process can be solved. Future research should answer the questions raised and indicate whether this process has a place in the production of hydrogen.

The Commission of European Communities Joint Research Center, Ispra (Varese), Italy, has a laboratory bench-scale thermochemical process operating on the following series of reactions:

$$SO_2 + Br_2 + 2H_2O \longrightarrow H_2SO_4 + 2HBr \quad (90^{\circ}C)$$

$$2HBr \longrightarrow H_2 + Br_2 \quad \text{(by electrolysis)}$$

$$H_2SO_4 \longrightarrow H_2O + SO_2 + \tfrac{1}{2}O_2 \quad (850^{\circ}C)$$

This process shares with the General Atomic Company process the high-temperature decomposition of sulfuric acid for the production of oxygen. It differs from the General Atomic process in using bromine in place of the iodine and in achieving the decomposition of the hydrobromic acid by electrochemical means rather than by pure thermochemical means. In a sense, this compromises the intention of the research directed at the thermochemical production of hydrogen. The goal of the thermochemical process investigator was to provide a process that would not require the generation of electric energy. This goal was desirable because it would eliminate the capital cost of the equipment necessary for the production, control, and handling of the electrical energy and would eliminate the inefficiencies of the heat to steam, to turbine, to electricity involved in the production of electric power.

The investigators at Ispra were attracted to this option because of the great difficulty in obtaining engineering materials from which to manufacture a plant that would stand up to hot halogen acids and free halogens such as bromine and iodine and their acids. The advantage to this type of hybrid process over the direct electrolysis of water is in the lower energy requirement for the electrolysis of the halogen acids. The electrolysis of hydrobromic acid takes only a little over half as much electrical energy as the electrolysis of water, and hydroiodic acid requires even less.

All the formidable problems of the high-temperature decomposition of sulfuric acid are still present with this hydrid process, but the problems of thermal decomposition of halogen acids is exchanged for the problems of halogen acid electrolyzers. It is the opinion of the workers at Ispra that these problems as a group are more manageable than the problems associated with the thermal decomposition of the acids. The Ispra laboratory plant has been operated for short periods of time during 1978 and 1979, and plans have been made for improving the plant to achieve continuous operation. Figure 4-16 shows the laboratory-scale pilot plant for demonstration of this process.[11]

The Westinghouse Electric Corporation has a laboratory bench-scale thermochemical hydrogen process based on these reactions:

$$2H_2O + SO_2 \longrightarrow H_2 + H_2SO_4 \quad \text{(by electrolysis)}$$

$$H_2SO_4 \longrightarrow H_2O + SO_2 + \tfrac{1}{2}O_2 \quad (870^\circ C)$$

The Westinghouse process, like the process developed at Ispra, is not a pure thermo-chemical process, and only part of the energy necessary to break down the water is supplied as heat; the remainder is supplied as electrical energy. This process has the advantage of eliminating the processes involving the corrosive halogens. It still retains, however, the high-temperature thermal decomposition of sulfuric acid for which no wholly satisfactory construction material has been identified for the decomposer[12].

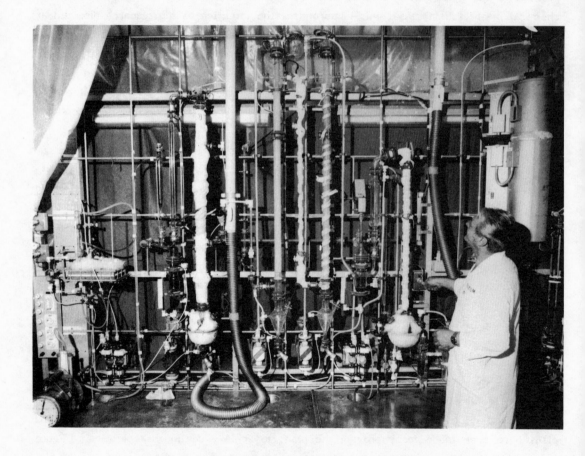

Figure 4-16. Ispra Sulfuric Acid Bromine Laboratory-Scale
Pilot Plant for the Production of Hydrogen
by Thermochemical Means (Courtesy Commission
of European Communities Joint Research Center,
21020 Ispra (Varese), Italy).

The promise of the thermochemical hydrogen production cycles is to identify a process that can take heat directly from a high-temperature source--fusion nuclear, fission nuclear, or solar--to drive the cycle with an overall system efficiency greater than that obtained with an electrolysis plant. In an electrolysis plant, the steam process by which the heat from the primary energy source is converted into electric energy is at best 40-percent efficient, and the electrolysis process is at best 90-percent efficient. This leads to an overall process efficiency from source heat to potential energy in product hydrogen of 36 percent. The initial projected theoretical efficiency of the thermochemical cycles under investigation is near 50 percent. If this efficiency can be obtained in practice with cycles that can be contained in plants constructed from available construction materials, the thermochemical production processes will become the mainstay of the hydrogen production industry. Thus far, however, it is not clear that there will ever be a competitive process. The theoretical promise has sustained a 15-year-long research effort involving a large number of investigators. The following organizations reported research activities at the Second World Hydrogen Energy Conference, Zurich, Switzerland, August 21-24, 1978:

Atomic Energy Commission of Israel
Beer-Cheva, Israel

Centre d'Etude de L'Energy Nucleaire
Mol, Belgium

Centre d'Etudes Nucleaires de Saclay
Gif-sur-Yvette, France

Deutsche Forschungs-und Versuchanstalt fur Luft-und Raumfahrt e.V.
Stuttgart, Federal Republic of Germany

Electric Power Research Institute
Palo Alto, California, U.S.A.

General Atomic Company
San Diego, California, U.S.A.

Institute of Gas Technology
Chicago, Illinois USA

Joint Research Center, Commission of European Communities
Ispra, Italy

Laboratories de Marcoussis
Marcoussis, France

Lawrence Livermore Laboratory
Livermore, California, U.S.A.

Los Alamos Scientific Laboratory
Los Alamos, New Mexico, U.S.A.

Mitsui Engineering and Shipbuilding Co., Ltd.
Chiba, Japan

National Chemical Laboratory for Industry
Tokyo, Japan

Nuclear Research Centre (KFA)
Julich, Federal Republic of Germany

Oak Ridge National Laboratory
Oak Ridge, Tennessee, U.S.A.

Rhein.-Westf. Technische Hochschule Aachen
Aachen, Federal Republic of Germany

University of Tokyo
Tokyo, Japan

Westinghouse Electric Corporation
Pittsburgh, Pennsylvania, U.S.A.

SOLAR PROCESSES

A few other processes for hydrogen production are under investigation. They are
all in the most early exploratory stages of study, so it is difficult to evaluate either

their ultimate technical or economic use compared with the existing processes. Both use sunlight as the source of energy for the decomposition of water. If the current development indicates feasibility, their long-term significance could be very great.

The process of photosynthesis carried out by plants results in hydrogen being taken from water and added to various carbon compounds. Under conditions where it is not possible for the plant to form the carbon-hydrogen compounds, the plant cells will release hydrogen directly. If a sufficient understanding of these reactions can be reached, it could be possible to separate the photosynthetic production of hydrogen from the rest of the biochemical machinery of the cell. This would provide a room-temperature reaction that would effect the breakdown of water into its component hydrogen and oxygen with the energy from the sun. There are formidable problems to be solved before this concept can lead to a process of commercial importance. The details of the photosynthesis process that is carried out by a living plant are today only beginning to be understood. Separating the many enzymes and intermediate products involved in the process, identifying them and understanding their function, and finally modifying them so that they can perform their task independent of the supporting biochemistry of the cell will take many years. Plants convert only a few percent of the incident sunlight in high-energy biochemicals, so it would be desirable to improve on the efficiency of the process. A process based on this concept may become important in the future but, for the next 10 to 20 years, it will be necessary to rely on other methods for the production of hydrogen.[14,15]

When light is absorbed by certain types of semiconductors, the energy absorbed appears in the form of positive- and negative-charged entities in the solid. The negative charge is carried by electrons that are freed from the individual atoms to wander freely through the crystal lattice of the solid. The positive charge resides in the hole left in the atomic structure of the crystal lattice by the removal of the electron. If a bound valence electron of a neighboring atom is captured by an existing hole, it leaves a new hole at its former position in the crystal lattice. When a sequence of these captures occurs, the hole appears to migrate through the crystal lattice as if it were an entity of real substance rather than just the space from which an electron is missing. Hole and electron pairs in ordinary semiconductors do not have enough energy to split a water molecule, but they do have enough to drive an electrical current through the semiconductor. Capturing this current is the basis of operation of the photovoltaic cells used to power spacecraft electrical systems or to provide modest amounts of electric power in other remote locations. The voltage of a single photovoltaic cell like the hole electron pair is not energetic enough to decompose water. Several of these cells can be placed in series to provide a voltage high enough to perform conventional electrolysis, but this would only be a different source of electricity for the conventional electrolysis cells that have already been discussed, not a new method of producing hydrogen. Several investigators have observed that, when certain types of semiconductors are illuminated while submersed in an electrolyte solution, hydrogen is evolved. This hydrogen cannot result from ordinary electrolysis of the water, for the voltage produced is not large enough to effect the breakdown of water. Some type of cooperative effect where several holes or electrons react in unison to reduce the water must be taking place.[16] The observed efficiencies of this process are quite low, 0.5 to 2 percent, but with research to explain the details of the process that occurs on the surface of the semiconductor, it may be possible to improve this to a value of importance for the production of hydrogen.

Chapter 4 References

1. M.J. Milner and D.M. Jones, "Hydrogen Production by Partial Oxidation," Proceedings of the 8th World Petroleum Congress, 1971.

2. H. Perry, "The Gasification of Coal," Scientific American, Vol. 230, No. 3, March 1974.

3. N.P. Cochran, "Oil and Gas from Coal," Scientific American, Vol. 234, No. 5, May 1976.

4. M.H. Miles, "Evaluation of Electrocatalysts for Water Electrolysis in Alkaline Solutions," J. Electroanalytical Chemistry, Vol. 60, 1975.

5. J.M. Gras and Le Coz, "Asbestos Corrosion Studies in Hot Caustic Potash Solutions," Proceedings of the 2nd World Hydrogen Energy Conference, Zurich, 1978, Pergamon Press.

6. J.H. Russell and L.J. Nuttall, "Development of Solid Polymer Electrolyte Water Electrolyzers for Large Scale Hydrogen Generation," CONF-781142, Proceedings of the U.S. Department of Energy Chemical/Hydrogen Energy Systems Contractor Review, 1978.

7. J.E. Funk and R.M. Reinstrom, "Energy Requirements in the Production of Hydrogen From Water," Industrial Engineering Chemistry Process Development, Vol. 5, No.3, 1966.

8. C. Marchetti, "Hydrogen Production from Water Using Nuclear Heat," Commission of the European Communities, Ispra, Italy, Code No. 1910825, 1970.

9.   R.D. Dafler et al., "Assessment of Thermochemical Hydrogen Production," Report No. COO-4434-14, Institute of Gas Technology for the U.S. Department of Energy, 1979.

10. J.R. Schuster et al., "Bench Scale Investigations and Process Engineering on the Sulfur-Iodine Cycle," CONF-781142, Proceedings of the U.S. Department of Energy Chemical/Hydrogen Energy Systems Contractor Review, 1978.

11. D. Van Velzen et al., "Development, Design and Operation of a Continuous Laboratory-Scale Plant for Hydrogen Production by the Mark-13 Cycle," Proceedings of the 2nd World Hydrogen Energy Conference, Zurich, 1978, Pergamon Press.

12. G.H. Parker, "Status Report on the Westinghouse Hydrogen Production Process," CONF-781142, Proceedings of the U.S. Department of Energy Chemical/Hydrogen Energy Systems Contractor Review, 1978.

13. T.N. Veziroglu and W. Seifritz, "Hydrogen Energy System: Proceedings of the 2nd World Hydrogen Energy Conference", 5 volumes, 1978, Pergamon Press.

14. T. Yagi and H. Ochiai, "Attempts to Produce Hydrogen by Coupling Hydrogenase and Chloroplast Photosystems," Proceedings of the 2nd World Hydrogen Energy Conference, Zurich, 1978, Pergamon Press.

15. A. Mitsui, "Bio Solar Hydrogen Production," Proceedings of the 2nd World Hydrogen Energy Conference, Zurich, 1978, Pergamon Press.

16. A.J. Nozik, "Hydrogen Generation Via Photoelectrolysis of Water - Recent Advances," Proceedings of the 2nd World Hydrogen Energy Conference, Zurich, 1978, Pergamon Press.

General Review Sources

Institute of Gas Technology, "Survey of Hydrogen Production and Utilization Methods," Final Report, NAS 8-30757, for the National Aeronautics and Space Administration, Huntsville, Alabama, 1975.

Srinivasan, S., F.J. Salzano, and A.R. Landgrebe, editors, Proceedings of the Symposium on Industrial Water Electrolysis, The Electrochemical Society, Inc., Princeton, New Jersey, 1978.

Chapter 5

HYDROGEN TRANSPORTATION AND STORAGE

Although hydrogen has a number of unique physical and chemical properties, the storage and transportation techniques needed for it are very much the same as those applied to the storage and transportation of natural gas (methane). Like natural gas, hydrogen can be pumped through pipelines, carried in high-pressure cylinders as compressed gas, or liquefied and transported as a liquid. It can be stored as a high- or low-pressure gas or as a liquid. Two new storage techniques that depend on the unique properties of hydrogen are under development. Both have shortcomings, but they may be used for special-purpose hydrogen storage: metallic hydride storage and microsphere storage. All the techniques used for natural gas are applicable to the transportation and storage of hydrogen, but all must take into account its very light weight.

Although hydrogen releases on combustion about 2.5 times the per-unit weight than natural gas, on a volumetric basis hydrogen has only about one-third the energy content. For a pipeline to carry the same amount of energy as hydrogen as it does carrying natural gas, about three times the volume of gas must be transmitted through the pipe. Hydrogen has a very low molecular weight and viscosity, factors that influence the rate at which a gas flows through a pipe under a given driving pressure. These two factors combine to give hydrogen about three times the velocity, at a given pressure drop, that can be achieved with natural gas. As a result, a given pipeline can carry about the same amount of energy in the form of hydrogen as in the form of natural gas. Unfortunately, the volume of gas is also three times as much, and the energy necessary to compress the gas to drive it through the pipe is almost entirely dependent on the volume of the gas being compressed. An established pipeline can deliver the same amount of energy carrying hydrogen as it does carrying natural gas, but the pumping energy will be nearly three times as much. When hydrogen is substituted for natural gas sometime in the future, the existing natural gas pipelines can be used to deliver the amount of energy they had in the past, but the pumping stations needed to keep the gas flowing will have to be increased in capacity.

The rate that a gas leaks through a small hole is like the rate of flow through a pipeline; it depends on the square root of the molecular weight. For hydrogen (molecular weight = 2) compared to natural gas (molecular weight = 16 (methane)), this ratio is as follows:

$$16^{1/2} \ / \ 2^{1/2} = 4/1.4142 = 2.8284$$

Hydrogen flows through a pipeline 2.8 times faster than methane, but it also leaks through imperfections or flaws in the pipeline 2.8 times faster than methane. Just as the pipeline carries about the same amount of energy whether it carries hydrogen or natural gas, the leakage of hydrogen, although 2.8 greater on a volumetric basis, is essentially the same on an energy basis.

Hydrogen is absorbed by some metals to form weak, brittle interstitial metallic hydrides. These compounds were discussed at length in Chapter 2, and more will be said about them later. When they form in materials used for construction in the transport or storage of hydrogen, they can weaken the structure of the metal and lead to failure. A number of research efforts are probing the nature of this problem, its importance with common construction materials, and methods of prevention. The preliminary results of these investigations indicate that there will be minimum problems from hydrogen embrittlement, but that further research is needed. Most common construction alloys do not form these hydrides to any significant extent.[1] When working with a metal that does form the hydrides, direct formation from hydrogen gas is very strongly inhibited by the presence of polar impurities in the hydrogen. In many cases, water vapor is an excellent inhibitor for the formation of the hydrides at levels as low as 100 parts per million. Hydrogen sulfide, carbon dioxide, alcohols, acetone, and similar compounds also show inhibition of the formation of the hydrides. The formation of these hydrides has been observed only when metals that are extremely clean are exposed to hydrogen that is very pure, in particular, free of these polar compounds. Although further work will be needed, it is expected that the existing natural gas pipeline network will be able to carry hydrogen as safely and reliably as it does natural gas, with essentially no problem from hydrogen embrittlement.

There are significant pipelines in operation that handle hydrogen gas primarily or exclusively. They are located in the refinery industrial complexes throughout the world. Several lines up to one kilometer or more in length exist in the refinery com-

plex near Houston, Texas.   These pipelines carry hydrogen from its place of manufacture by steam reforming of hydrocarbons to ammonia synthesis plants or desulfurization
facilities for crude oil.   In the Ruhr Valley between Germany and France, there is a
hydrogen pipeline about 30 kilometers long that has operated without incident for over
30 years.

Test programs are underway to define the boundaries of the use of hydrogen in
existing natural gas equipment.   The Institute of Gas Technology has a laboratory flow
loop system, containing equipment currently designed for handling natural gas, under
continuous exposure to flowing hydrogen.   This test is illustrated in Figure 5-1.   Sandia
Laboratories has in operation a high-pressure flow loop for the testing of pipeline structure and fabrication techniques during exposure to pure hydrogen.   This flow loop is
shown in Figure 5-2.

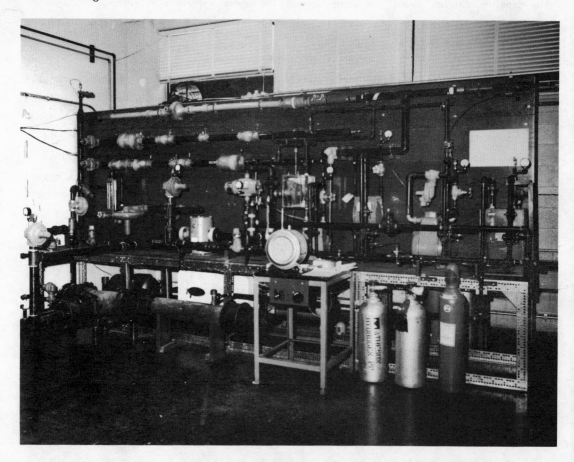

Figure 5-1.   IGT Test Flow Loop for Evaluation of Natural
Gas Equipment in Exposure to Flowing Hydrogen
(Courtesy of The Institute of Gas Technology
(IGT), 3424 South State Street, Chicago,
Illinois 60616, U.S.A.).

Figure 5-2. High-Pressure Hydrogen Flow Test Loop
Evaluating Pipeline Construction Materials
and Techniques (Courtesy of Sandia Laboratories,
P.O. Box 5800, Albuquerque, New Mexico 87115,
U.S.A.).

Liquid hydrogen can be transported through pipelines, but these lines must be extremely well insulated. Short lengths of vacuum-insulated pipelines for the transport of liquid hydrogen have been used for a number of years in the space program and in liquid hydrogen production plants. These lines consist of a center pipe for carrying the hydrogen, surrounded by a 2- to 5-centimeter layer of aluminized mylar plastic, each layer separated from the others by a layer of nylon net. An outer pipe encloses the mylar superinsulation and forms a vacuum-tight container for it. The insulation layer is evacuated to a good quality vacuum (better than 10^{-4} torr). This type of rigid pipe will transmit liquid hydrogen for long distances without undue loss from boiling of the cold liquid hydrogen, but it is quite expensive and its use for transporting liquid hydrogen any great distance (beyond a few hundred meters) has not proven cost-effective.

Using a technique similar to the inflexible pipelines, it has been possible to fabricate flexible liquid hydrogen pipelines for filling over-the-road trailers and tank cars. These flexible pipelines are made from two flexible, concentric metal bellows. The inner bellows is 2 to 5 centimeters smaller in diameter than the outer and is surrounded at intervals by rings that prevent the two bellows from touching. The space between the rings is filled with insulation. These flexible lines demonstrate a higher heat leak than rigid lines, but are adequate for short connections between stationary and mobile tanks. As a consequence of the very low temperature and thus low reactivity, liquid hydrogen does not present any problem of hydrogen embrittlement, but some common construction materials become brittle from the cold alone. Fortunately, a number of common aluminum alloys, low-carbon steel, and stainless steels have adequate ductility at the temperatures of liquid hydrogen for construction purposes. There is therefore no problem with materials being unavailable or priced too high for the construction of liquid hydrogen handling equipment.

Hydrogen is transported as a high-pressure gas in cylinders at pressures ranging from 150 to 400 atmospheres (Figure 5-3), a technique that is well developed, quite

Figure 5-3.   A Tube Trailer for the Shipment of Modest
Amounts of Hydrogen (Courtesy of Air Products
Company, Box 538, Allentown, Pennsylvania 18105,
U.S.A.).

reliable, and convenient, but very inefficient. A steel cylinder weighing 20 to 30 kilograms will hold only 1 kilogram of hydrogen at 150 atmospheres pressure and 2.5 kilograms at 400 atmospheres. This great container weight results in a transportation system in which only 2 to 4 percent of the weight of the cargo is the desired material. The cost of shipping generally depends on the weight of the shipment, such that the cost of shipping hydrogen in cylinders is very high. This technique is used for the delivery of small amounts of gas for applications in which the actual cost of the hydrogen is not very important, but it would be too expensive for the transport of hydrogen for use as vehicle fuel. For some applications, particularly private passenger vehicular hydrogen storage, there are serious doubts concerning the safety of high-pressure gas storage. If a fully charged cylinder at 150 atmospheres pressure were to rupture in an accident, the effect would be similar to an explosion. Much damage would be done to the vehicle and occupants beyond the direct damage of the accident. This effect would, of course, be more extreme for tanks filled at higher pressures, but higher pressures are desirable because they improve the efficiency of the containment method. The necessary ancillary equipment--valves, regulators, flow meters, relief valves, and pressure gauges--for the handling and transfer of hydrogen gas from pipelines and transfer cylinders is developed and available.

As with transport, hydrogen can be stored in much the same manner as natural gas. High-pressure gas storage is currently the most common method of sorting hydrogen in modest quantities. Cylinders can be used for short- or long-term storage or, as pointed out above, can be used as a storage medium for hydrogen transport. This technqiue is most satisfactory in stationary applications where the great weight of the storage container as compared to the weight of the stored hydrogen is not relevant.

Research and development is underway leading to the adoption of some of the recently developed ultrastrong filaments--graphite, glass, boron, amorphous carbon, and organic polymers such as Kevlar®--to the manufacture of lighter weight cylinders for the containment of high-pressure gas. These developments, if successful, will lead to cylinders that are as much as one-fifth to one-tenth the weight of the current steel cylinders. This will have a significant effect on the use of high-pressure gas as a transportation medium, but because these new technology cylinders will undoubtedly be more expensive, they are not expected to replace the heavier but less costly steel cylinders in static storage applications.

In all high-pressure storage techniques, the cost to compress the gas must be accounted for in the selection of the method. Energy is required to compress gas. On initial compression, the gas becomes hotter as well as compressed. If the gas is allowed to cool, the energy represented by the higher temperature immediately after compression is lost, and only the pressure component of the energy can be recovered. In many applications, it is difficult to recover even the pressure energy. The cost of storing hydrogen at high pressure always carries the burden of the cost of compressing the gas.

In the past, when coal gas was used for home service purposes, it was common to see large, low-pressure gas holders. These gas holders operated like a drinking glass inverted in a pan of water. As gas was passed into the container, the tank would float higher and higher. The only pressure to be overcome was that necessary to lift the large, lightweight container. The water seal at the bottom prevented the gas from leaking out and, at the same time, allowed the tank to respond to the inflow or out-flow of gas in essentially a frictionless manner. These tanks were safe and reliable but were quite large. To hold enough hydrogen for one winter of heating for a town of 100,000 would require a low-pressure cylindrical gas holder about 300 meters high and 300 meters in diameter. Such a structure is probably technically feasible to construct, but the citizens of a town might find it more of a landmark than they could tolerate. Of course, the storage could be divided into many smaller tanks or periodic resupply could be arranged. In the early days of this century, this method of gas storage was the most practical for urban uses including cooking, home heating, and light industrial use, and this is probably true today. It is also one of the safest methods of gas storage. There are no high pressures to rupture tanks or pipelines or to drive large volumes of gas out of the tank. Because of the lack of driving pressure, leaks in the tank would not immediately result in the loss of all the stored gas; a complete loss would take a significant length of time. In the event of a large hole, split, or rupture in the tank, probably the most serious incident, followed by ignition of the gas, there would be a very large fire but no explosion. In the future, when hydrogen power be-comes the rule, these tanks may again be seen in our cities.

Hydrogen can also be stored as a liquid (Figure 5-4), with the major considera-tion being the requirement for insulation. Hydrogen has a boiling point of only 20.38°K (see Chapter 2) and a heat of vaporization of only 446 joules per gram, factors contri-buting to a rapid rate of boiling from the smallest leakage of heat into the storage vessel. In some applications, for example, the liquid hydrogen tanks used in the upper stages of the Saturn V launch vehicle and the drop tank used on the Space Shuttle, a 10- to 15-centimeter layer of plastic foam can be used. These applications do not re-quire very high quality insulation. The vehicle is placed on the pad, and the liquid hydrogen tank is filled only an hour or so before lift-off. During the time it sits on the pad before actual lift-off, the liquid hydrogen tank is continuously vented and

Figure 5-4.   Liquid Hydrogen Storage Tanks (Courtesy of Air
              Products Company, Box 538, Allentown,
              Pennsylvania 18105, U.S.A.).

topped-off to maintain the proper fuel load. During the flight, the hydrogen is used so rapidly (completely used in 4 to 5 minutes) that there is essentially no effect from the slight amount of boiloff that occurs.

Plastic foam insulation will probably be suitable for use in hydrogen-fueled airplanes, for much the same reason that it is adequate for space launch vehicles. The airplane will be fueled just before takeoff, and even though the boiloff rate is high, the use rate is even higher. Therefore, in addition to withdrawing the boiloff vapor for fueling the engines, liquid will also have to be withdrawn and vaporized in a heat exchanger to provide the fuel required to power the airplane.

In both these aerospace applications, the weight of the insulation is more important than the quality, once a minimum quality is obtained. At first it might seem possible to get along without insulation, since rapid boiloff is acceptable from a fuel-use viewpoint. Insulation is necessary, however, because liquid hydrogen is so cold that, in a bare metal container, it will condense the air surrounding it to a liquid and build up very thick frost layers with great rapidity. The condensation of air initially produces a mixture of liquid nitrogen and oxygen, but as a consequence of the higher boiling point of oxygen than nitrogen, the liquid rapidly becomes enriched in liquid oxygen. If this liquid oxygen were to dribble onto combustible materials on or near the launch pad, an explosion could result. In addition, the frost buildup can alter the aerodynamic shape of the vehicle and add enough weight that the vehicle no longer will perform properly.

For longer storage of liquid hydrogen, it is necessary to use vacuum insulation, the best being termed vacuum superinsulation. Heat can be conducted across the gap between the outer vacuum barrier and the inner liquid hydrogen tank by two mechanisms: conduction and radiation. If a gas were allowed to remain in the gap, the gas near the outer shell would absorb heat through the outer shell from the sur-roundings. This heat would be conducted to the next layer of gas until it was finally absorbed by the inner tank. Evacuating the space between the inner and outer shells essentially stops this conduction of heat. Heat can also be transmitted across the gap by radiation. In superinsulation, a number of layers of mylar plastic coated with aluminum are interposed between the two shells. Much of the radiation from the outer shell is reflected back to the outer shell by the first layer of aluminized mylar, that which is transmitted by the first mylar layer is reflected back by the second mylar layer, and so on until almost no radiation reaches the inner shell containing the liquid hydrogen. In a tank designed in this manner, very little heat reaches the inner tank directly through the walls, and most of the heat leak is through the supports that must be provided for the inner tank and the fill and drain lines used for placing the liquid hydrogen in the tank and removing it for use. The best quality tanks manufactured in this manner suffer a boiloff rate of about 1 percent per day in sizes of 100 to 400 liters. Larger tanks have a lower rate of loss.

The thickness of the superinsulation layer ranges from 2 to 5 centimeters, with as many as 100 layers of aluminized mylar, each separated by a lightweight nylon net to reduce the thermal conductivity through the mylar layers. To perform in the optimum manner, this insulation must be very carefully layered and cannot have holes or com-pressed areas. All these tanks are presently made by careful hand layup of the layers, with great care being taken to ensure uniformity of coverage. As a result of the large amount of hand labor involved in their manufacture, these tanks are relatively expensive and cannot be fabricated in very large sizes.[2]

Large liquid hydrogen tanks use a vacuum insulation called vacuum perlite, which does not have quite the low heat leak of superinsulation but which is much cheaper to make. There are again two shells, but they have a separation of 10 to 30 centimeters, and the gap is filled with perlite, an expanded mica. Perlite is produced by heating a selected variety of mica to 150°C. Water trapped in the mica boils and expands, leaving a silver-appearing inorganic structure with many layers of mica separated slightly from each other. The perlite is simply poured into the space between the two shells, sealed off, and evacuated. During the evacuation, the tank is heated to help drive off any volatile matter that may be present in the space to be evacuated. Large storage tanks using vacuum perlite insulation have been built for storing liquid hydrogen at production facilities and space launch sites. These tanks demonstrate boiloff rates well under 0.5 percent per day. The vacuum perlite insulation allows 5 to 10 times as much heat leak as does high-quality superinsulation but is much cheaper and can be applied in thick layers to very large-sized tanks.[2]

Both liquid hydrogen and natural gas storage have a problem of liquid thermal stratification when they are stored in very large tanks. The liquid at the bottom of the tank is under a higher pressure than that at the top of the tank because of the hydrostatic pressure exerted by the column of liquid. The higher pressure at the bottom raises the boiling point of the liquid to a temperature somewhat above that at the surface. As time passes, two layers develop in the tank--a cold, relatively low-vapor pressure layer on top and a somewhat warmer, higher vapor pressure layer at the bottom. The cold top layer, as a result of its lower temperature, is somewhat denser than the layer on the bottom of the tank. This is obviously an unstable condition. If the tank is disturbed, the two layers roll over, bringing the slightly warmer but significantly higher vapor pressure liquid to the top of the tank. When this happens, there is an immediate strong rise in the pressure in the tank and a surge of boiling in the liquid. This phenomenon caused the failure of some of the early tanks designed for cryogenic storage; the vent lines were not sized large enough to handle the surge of vapor, and the pressure became high enough to rupture the tank. In very large tanks, this process is prevented by stirring the liquid slowly to prevent the thermal stratification. In smaller tanks, the problem has been solved in a passive manner by very loosely filling the tank with aluminum wool. The thermal conductivity of the aluminum wool is great enough to prevent the thermal stratification. Only a small amount of wool is required, small enough that it occupies only about 1 percent of the volume of the tank.

Vacuum-insulated liquid hydrogen tanks are in service in a large number of places throughout the world. They are used for storing many liquefied gases as well as liquid hydrogen and are used in over-the-road transporters of liquid oxygen, nitrogen, argon, helium, and fluorine as well as liquid hydrogen. Railroad cars have been equipped with vacuum-insulated containers for shipment of the same commodities. Liquid helium, which has an even lower boiling point than hydrogen (4.2°K), has been shipped across oceans in this type of storage vessel. There are currently several large ocean-going ships equipped with vacuum-insulated tanks of very large size that carry liquefied natural gas from Algeria to the East Coast of the United States. When hydrogen becomes the universal chemical fuel, the technology necessary to handle liquid hydrogen will be ready and waiting.

Although it is quite feasible and within the developed state-of-the-art to handle, store, and ship hydrogen as a liquid, a distinct penalty is incurred if this technique is used. Much energy is expended in providing the refrigeration necessary to liquefy hydrogen. The amount expended is at least equal to one-third of the total energy present in the liquid when it is burned. There are applications in the fueling of airplanes and automobiles where this stored negative energy can be put to use to great advantage, but for simple storage in a pipeline hydrogen distribution network, the penalty is probably too large and methods other than liquid storage will be used.

There are two new methods of storing hydrogen that are attracting much interest: metallic hydrides and microballoons. As discussed in Chapter 2, a number of metals reversibly absorb hydrogen into their crystal structure. If one of the hydrides that absorbs hydrogen at room temperature is placed in a gas-tight containment vessel, hydrogen can be pumped in and the heat of formation of the hydride removed. When charging is completed, the metallic hydride contains hydrogen at atom densities near or greater than that of liquid hydrogen. The pressure in the container is only slightly above atmospheric. To recover the hydrogen, the hydride must be heated, the pressure goes up, and the hydrogen is desorbed when an amount of heat roughly equal to the heat of formation is supplied back to the hydride. This technique has the promise of providing an excellent method of storing hydrogen in fixed locations. To store hydrogen for a city of 100,000 at essentially ambient pressure, a hydride container capable of holding the same amount of hydrogen as the gas holder discussed earlier in this chapter would be only about 30 meters high by 30 meters in diameter.

There are, however, still many development problems with the hydrides. The hydrides hold only 2 to 4 percent hydrogen on a weight basis, so for a 1-kilogram charge of hydrogen, a minimum of 25 kilograms of hydride material is required. While charging, the heat of hydride formation must be removed from the bed of solid particles; while discharging, heat must be supplied to the same bed. Heat transfer in and out of solids is difficult to achieve when the heat transfer medium must remain separated from the solid. The current technique is to pack the hydride into a shell and tube heat exchanger of high tube density, so that the separation distance of the hydride particle furthest from the heat exchange tube is kept as small as possible. This arrangement has produced acceptable performance but has added the weight of the complex heat exchanger to the already high weight of the hydride. The need to add and remove heat during the charge and discharge cycles requires the storage facility to have a heating and cooling facility or system to handle this heat exchange. The hydride material expands and flakes apart when the hydrogen is added. This effect continuously grinds and shifts the hydride as it is cycled through charge and discharge. Ultimately, this leads to packing of the hydride and reduced ability of the input hydrogen to flow in and out of the hydride storage bed.

The best alloy from the standpoint of ease of charge and discharge and pressure versus temperature is an expensive alloy of lanthanum and nickel ($LaNi_5$). An iron titanium alloy (FeTi) is less expensive but requires more heat to drive the hydrogen from its hydride. A new alloy of magnesium with various amounts of nickel has the promise of achieving the 4 percent by weight level of hydrogen containment, but it is a difficult alloy to formulate and requires an even higher temperature for the release of the hydrogen than does the iron titanium alloy.

Many of these hydrides can be easily poisoned. Trace amounts of polar gas such as water vapor, carbon dioxide, sulfur compounds, and polar organics react somehow with the surface of the hydride particles and greatly reduce their ability to absorb hydrogen. This can take place as a single event, when the hydride is exposed to these gases at levels of 500 to 1000 parts per million, or it can take place slowly, when hydrogen containing these kinds of impurities at levels of a few parts per million is cycled in and out of the hydride tank. Once poisoned, the hydride must be subjected to very high temperature and high vacuum to remove the impurities. Problems from poisoning can be avoided by using ultra-pure hydrogen in the charging, but this is not always economical or technically feasible in real-world systems. The hydrides at their current state of development are useful for certain limited applications, and with further research it may be possible to use them for the static storage of hydrogen to great advantage.[3]

Microballoon storage has some characteristics in common with the hydrides. The hydrogen is stored in a solid and is released by heating the solid. Glass manufacturers have developed a technqiue for the production of glass balloons or bubbles smaller than

0.1 millimeter. These balloons have very thin walls and, when heated to temperatures of 300° to 400°C, the glass becomes permeable to hydrogen. To charge these balloons with hydrogen, they are placed in a sealed chamber capable of withstanding high pressures at 300° to 400°C. The chamber is heated to the charging temperature and slowly pressurized to 800 atmospheres with hydrogen. After the full charging pressure is reached, the chamber is allowed to cool to room temperature. Each small balloon is charged with hydrogen at the charging pressure less the pressure drop due to the lower temperature. Pressures in the range of 400 atmospheres are possible in the cooled balloons. At these pressures, 5 to 10 percent of the weight of the charged balloons is hydrogen. To recover the hydrogen for later use, the balloons are heated again and the hydrogen diffuses back out. As with all new techniques, there are problems that must be worked out. Glass is tremendously strong when it is free of flaws and has no internal strain from cooling nonuniformly. Perfect 0.1-millimeter spheres that are free from flaws and strain can easily support pressures of 400 atmospheres, but they really must be perfect. During the initial charging operation, 60 to 90 percent of the balloons fail and crumble and must be separated from the charged balloons; the remaining charged balloons must be handled in such a manner that they are not broken. This technique, like the use of metallic hydrides, may be developed into a method for the storage of hydrogen, but much further research must be performed.[4]

It may be possible in the distant future to store hydrogen in the metallic form. If the metallic form could be produced, it would be an ideal method of hydrogen storage. Theoretical predictions indicate that metallic hydrogen might have a density of 1 to 2 grams per cubic centimeter, a density far exceeding that of any other storage technqiue. There has been no unequivocal experimental detection of metallic hydrogen, but best estimates indicate that it should be the stable form for hydrogen at pressures over 2 million (2×10^6) atmospheres. If it can be produced and is metastable at room temperature and pressure, it would provide the best method of storing hydrogen.

Chapter 5 References

1. H.R. Gray, American Society for Testing Materials, Special Technical Publication, STP 543, 1974.

2. C.R. Lemons, C.R. Watts, and O.K. Salmassy, "Development of Advanced Materials for Use As Insulation of Liquid Hydrogen Tanks," National Aeronautics and Space Administration Report, NASA-CR-123928, 1972.

3.   A.F. Andresen, and A.J. Maeland, editors, "Hydrides for Energy Storage," Pergamon Press, 1978.

4.   R.J. Teitel, T.M. Henderson, and J.E. Luderer, "Microcavity Systems for Automobile Applications," <u>Proceedings of the U.S. Department of Energy Chemical/Hydrogen Systems Contractor Review</u>, 1978.

General Review Sources

Johnson, D.G. and W.J. Jasionowski, "Study of the Behavior of Gas Distribution Equipment in Hydrogen Service," Institute of Gas Technology for the U.S. Department of Energy, EY-76-C-02-2907, 1979.

Phadke, L.G., et al., "An Assessment of Hydrogen Compressor Technology for Energy Storage and Transmission Systems," Clean Energy Research Institute, University of Miami for U.S. Department of Energy, EC-77-S-05-5598, 1979.

Chapter 6

HYDROGEN AS AN ENERGY CARRIER

There are a number of alternative primary energy sources that are candidates to help fill our future energy needs. The following table lists those that have any likelihood of being used in the near future to replace dwindling and environmentally harmful fossil fuel supplies.

Potential Energy Sources

Renewable Sources	Potential Uses
Solar	Space heat, industrial process heat
	Electric power generation
Wind	Electric power generation
Water	Electric power generation
Geothermal	Space heat, industrial process heat
	Electric power generation
Ocean tides	Electric power generation
Ocean thermal gradients	Electric power generation
Ocean waves	Electric power generation

Nonrenewable Sources with Long Lifetimes	
Thermonuclear fusion	Space heat, industrial process heat
	Electric power generation
Breeder fission nuclear	Space heat, industrial process heat
	Electric power generaton

Nonrenewable Sources with Short Lifetimes

Burner fission nuclear   Space heat, industrial process heat

                                              Electric power generation

All of these sources have a common shortcoming: they do <u>not</u> yield a storable or portable energy carrier. For the production or harvest of energy, most require fixed facilities that are massive in volume and extension. Their product is heat or mechanical energy that can be converted to electricity, but whatever the direct product, the output must be used immediately. There is no product that can be stored for later use.

All uses of energy can be grouped into one of four categories:

Use	Approximate Part of Total Energy Use	Examples
Heating space	1/4	Homes, buildings, factories
Heating things	1/4	Steel manufacture, chemical processes
Making machinery function	1/4	Factories, materials handling
Transportation	1/4	Moving people and things

The first three of these uses can be satisfied by the output from the potential alternative energy sources when there is an adequate match in both time and space between the energy source and the end use. For these uses, a storable energy form would be desirable to match production with use in time and a transportable form would be desirable to match production with use in space. For transportation use, a portable source of stored energy is absolutely necessary, and none of the primary energy sources available to us in the future meets this requirement.

The requirement for a material that can store energy for future use, provide a simple method for energy transport, and serve as a fuel for the transportation sector of the economy can be most easily met by hydrogen. A portion of the thermal energy produced by the power sources can be diverted to the direct production of hydrogen by a thermochemical cycle or converted to electricity for the production of hydrogen by electrolysis. Because no thermochemical cycles have been demonstrated to be technically feasible and as electrolysis is an established, successful industry, this discussion will assume the use of electrolytically produced hydrogen. The advent of a successful thermochemical cycle will in no way change the outcome of the large-scale use of hydrogen as a fuel.

Electrolysis is a simple electrochemical reaction in which a low-voltage electric current is passed through water that is conductive due to the presence of dissolved electrolytes (see Chapter 4). The most common electrolytes used are potassium and sodium hydroxide. Two electrodes are required for contact with the water. As a low-voltage direct current is passed by the electrodes through the water, oxygen is produced at the anode (positive pole) and hydrogen is produced at the cathode (negative pole). This process is nearly 100-percent efficient in the laboratory and can probably be performed at above an 85-percent efficiency in large-scale electrolysis of water. This high efficiency means that most of the electrical energy expanded to electrolyze the water is stored as potential chemical energy in the hydrogen produced. Two volumes of hydrogen are produced for each volume of oxygen. Although the hydrogen is produced at a high level of purity, it is saturated with water vapor. For practical transmission through pipelines, this water vapor must be removed from the hydrogen gas stream, because condensation of water in pipelines can cause problems in metering and valving, and can reduce the lifetime of the pipeline by promoting corrosion. The water can be removed by cooling the hydrogen until the water condenses out. This cooling can be accomplished by mechanical coolers or by allowing the hydrogen to expand from a high-pressure level to a much lower one, causing cooling by expansion. The water recovered by this process is distilled water of high quality and could be used for special purposes.

It is desirable to have the hydrogen produced at a high pressure for pipeline transport and to allow cooling by expansion to remove the water. Current technology electrolyzers produce hydrogen at a pressure only slightly above ambient pressure, so it would be desirable to develop electrolyzer processes that operate at high pressure. In high-pressure electrolysis, the energy required to compress the gas shows up as a slightly greater power requirement for electrolysis of the water. Because this compressive process is, like the electrolysis, extremely efficient, minimum energy is lost in producing the high-pressure gas. If the electrolysis is performed at pressures of 100 to 300 atmospheres, the pressure will be high enough to allow expansive cooling for water removal and still retain a high enough pressure (8 to 40 atmospheres) for satisfactory distribution. The cold, dry hydrogen produced by the water removal unit can be used for the other cooling needs of the electrolysis plant. This gas can be piped directly from the water recovery unit to the distribution network for transfer throughout the country without any intermediate pumping. Natural gas pipelines currently cover a large part of the United States and there are many in Europe. Many of these lines could be used without modification to transport hydrogen gas throughout the country;

others would require renovation to reduce leakage. In addition, new, separate pipelines could be built where desirable to handle the oxygen produced.

The cost of electric power generation by any type of power plant decreases with increasing generating capacity. This makes it desirable to build power plants as large as possible but not so large that the cost of distribution of the power becomes prohibitive. With electric power, there are definite limits to the distance the power can be transmitted. This limiting distance also limits the total number of customers who can be served by a single power plant, and in turn limits the size of the power plant itself. Even the most advanced ultra-high-voltage lines are limited to about 500 kilometers for economic power transmission.[1] Practical transmission lines are seldom more than 150 to 300 kilometers long before resistance and high-voltage corona losses (leakage of high voltage directly into the air) make them uneconomical. These transmission losses place a practical upper limit on the size of electrical generating plants.

On the other hand, gas transmission shows only low losses over continental distances. Conversion of the electric power to potential chemical hydrogen power by electrolysis of water lowers one of the main barriers to the use of very large power generating plants because it allows potential low-cost power (the hydrogen gas) to be transmitted over long distances with low losses. Unlike conventional electric plants that must vary their output over a large ratio to meet the variable demands throughout the day, the electrolysis power plant can operate at a constant output level. This advantage arises from the ease with which the hydrogen can be stored as a gas or liquid at the generation site or near the user. Demand-smoothing storage can be obtained at the generating site simply by using the old technology of gas holders or pressure cylinders and possibly the new metallic hydrides. For longer term storage, for example, during the summer months, the hydrogen can be liquefied and stored in underground cryogenic vessels as is done today with liquid natural gas. Hydrogen boils at -252°C, compared to -161°C for natural gas. Although this 91°C temperature difference will make insulation problems more severe with hydrogen, it will not create any new requirements. New liquefaction techniques developed in conjunction with the space program have greatly reduced the cost of liquefying gases.[2] Of course, at locations far from the power plant, the power required to liquefy the gas can be obtained from the hydrogen fuel itself.

HYDROGEN AS A FUEL

Along with the advantage of being able to transmit the hydrogen gas over long distances with low losses is the extreme value of the portability of hydrogen. This portability can be obtained by liquefaction of the hydrogen for use as the fuel in most mobile systems.   The liquid hydrogen used in mobile power generation can be produced in the same liquefaction facilities that are used to produce liquid for long-term storage.   For mobile power use, it may be desirable to use the slush hydrogen technique developed by the National Bureau of Standards.   In this technique, the hydrogen is cooled until part of it solidifies; the solid is then broken into small particles and stirred into the liquid to produce a liquid hydrogen/solid hydrogen slush.[3]   This produces up to a 15-percent increase in the density of the stored fuel and improves the storage life, for the solid hydrogen absorbs a significant amount of heat without a temperature change when it melts.   As either liquid or slush, hydrogen should fill the requirement for a portable zero-pollution fuel.

The hydrogen fuel made available by electrolysis will not pollute the atmosphere and hydrosphere.   Hydrogen gas burns in air to produce only water.   No carbon dioxide, carbon monoxide, or carbon soot is possible; no poisonous organic compounds are formed; and no particulate ash remains when the only fuel used is hydrogen gas.   The burning of hydrogen gas would produce no sulfur dioxide, as produced in large quantities by burning soft coal and low-grade fuel oil in power plants.

The oxides of nitrogen produced by high-temperature combustion processes can be suppressed by increasing the fuel used in the combustion process.   With conventional fuels, this fuel-rich combustion eliminates the oxides of nitrogen, but it causes a large increase in the production of carbon monoxide and organic chemical residues.   When hydrogen is used as the fuel in fuel-rich combustion, the oxides of nitrogen can be eliminated with an attendant release of only a small amount of nonpolluting hydrogen gas.   The product of hydrogen combustion--high-purity water that is nonpolluting--could be collected by condensation and used to augment existing water supplies where feasible and desirable.   Theoretical equilibrium calculations are shown in Figures 6-1 and 6-2.

For these calculations, a mixture of air and fuel is assumed, and the equilibrium chemical mixture is calculated at the temperature of the combustion.   In the fuel-lean area of Figure 6-1, the amount of nitrogen oxide produced by the combustion of hydrogen is somewhat higher than that produced by the combustion of gasoline.   This occurs because the flame temperature of the hydrogen combustion is higher than that of gasoline combustion.   High flame temperatures favor the formation of oxides of nitrogen when there is excess oxygen present to combine with the nitrogen of the air.   In the fuel-rich area of the curve, the oxides of nitrogen are strongly suppressed because the equilibrium favors the formation of the oxidation products of the fuel components rather than the oxidation of the nitrogen.   In the fuel-rich portion of Figure 6-2, the presence of carbon monoxide (CO) and unburned hydrocarbons (CH_4 and C_2H_4) becomes greater and greater as the concentration of nitrogen oxides decreases.   It is interesting

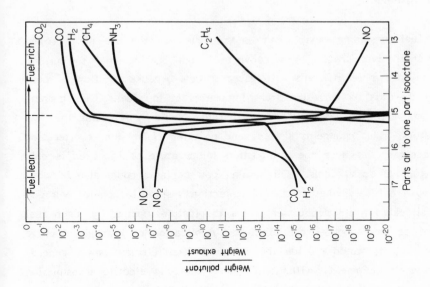

Figure 6-2.

Theoretical Calculations showing the Chemical
Equilibrium Exhaust Products of the Combustion
of Gasoline (iso-octane).

NO = Nitric oxide	H_2 = Hydrogen
NO_2 = Nitrogen dioxide	CH_4 = Methane
CO = Carbon monoxide	NH_3 = Ammonia
CO_2 = Carbon dioxide	C_2H_4 = Ethylene

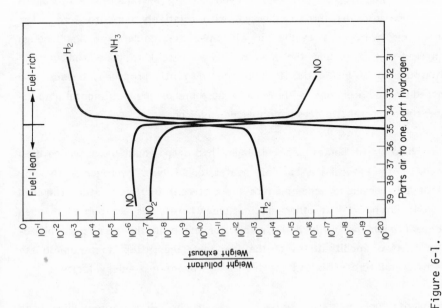

Figure 6-1.

Theoretical Calculations Showing the Chemical
Equilibrium Exhaust Products of the Combustion
of Hydrogen.

NO = Nitric oxide	H_2 = Hydrogen
NO_2 = Nitrogen dioxide	NH_3 = Ammonia

to observe that, in the fuel-rich combustion zone, the amount of hydrogen and ammonia produced by the combustion of hydrogen is about the same as that produced by the combustion of gasoline.

Every existing fuel-consuming device can be converted to the use of hydrogen gas. Large powerplants have already been converted from burning coal to burning natural gas and are in fact more efficient with gaseous fuels because the flow is more easily controlled and there is no corrosion from impurities in the fuel. Coal-burning power plants can be converted to gaseous hydrogen-burning plants by replacing the coal grates with gas injectors. This change is simple, and no other conversion or alteration of the furnace is necessary, because the combustion temperature of hydrogen is only slightly higher than that of fossil fuels. Oil-burning power plants could also be converted to hydrogen with only a change in the oil injection system to handle hydrogen gas. Conversion of all existing electric power plants to pipeline-supplied hydrogen fuel would eliminate them as local sources of pollution. Application of hydrogen fuel to existing electric power plants would provide the means of local electric power generation with existing capital equipment, without the need for long electric transmission lines from the very large future primary power sources located in remote places around the world.

Industrial processes that use coal and oil-based fuels can also be converted to burn hydrogen. For example, blast furnaces used in the production of iron from iron ore could be converted from coke burning to hydrogen burning by injecting a hydrogen-rich hydrogen/air mixture into the furnace charged with relatively pure iron ore. The partial combustion of the hydrogen with the air will raise the temperature high enough that the excess hydrogen will reduce the iron ore (iron oxide) to pure iron metal.[4] The resultant iron would be of higher quality than that currently produced, because the coke now used to reduce the iron ore is the source of some of the detrimental impurities like silicon and sulfur present in freshly produced pig-iron.

For use in the heating of homes and buildings, hydrogen would be a choice fuel because it is clean. Any natural-gas-burning heater could burn hydrogen with only minor adjustments of the burner to accommodate the different fuel/air mixture required for efficient hydrogen burning. Most other applications where fuels are burned in open flames to produce heat could be easily and cheaply converted to the burning of hydrogen by minor adjustments or modifications of the fuel-metering system to accommodate the different fuel-to-air mixture ratio and the higher hydrogen/air flame velocity.

Mobile power plants such as those used in trains, trucks, ships, automobiles, and airplanes are somewhat more difficult to convert to hydrogen use than are stationary

plants.   Current technology is available (Figure 6-3), with development, to convert
these burners of fuel to combustion of hydrogen gas.   The conversion will require the
application of the techniques of handling cryogenic liquids learned from current space
programs.   For bulky vehicles, such as trains and ships, it would be possible to install
currently available cryogenic storage vessels to contain liquid hydrogen to be used as
the fuel.

Figure 6-3.   Air Products' Two 30-ton-per-day Liquid
Hydrogen Plants at the Company's Site in
New Orleans, Louisiana (Courtesy Air Products
Company, Box 538, Allentown, Pennsylvania 18105,
U.S.A.).

External combustion steam turbine engines and gas turbines would require very simple modifications because the fuel feed and combustions are continuous. As with the gas-fired stationary power plant, only the fuel injection jet size would need to be changed to match the different hydrogen-to-air mixture ratio.[5]

THE AUTOMOBILE

Work done to date indicates that the conventional internal combustion engine can, with minor modification, be operated with hydrogen as the fuel. Well-developed methane and propane carburetors are now widely used on automobile-type engines. Although their flow-control orifices are not suitable for use with hydrogen because of the different volumetric ratio required, redesign would be simple. The ancillary pressure regulators, valves, controls, etc., are also available for these carburetors, and they too would require only a modest amount of redesign or resizing to handle hydrogen.

Hydrogen has a tendency to knock and backfire when it is pre-mixed with air and fed to an engine with a compression ratio of greater than 7 to 1. This problem is most severe when the fuel-to-air mixture ratio is near stoichiometric, which is the mixture that will produce the maximum power.[6] The knock can be suppressed by operating at a lean fuel-to-air mixture or by diluting the charge with exhaust gas recirculation or water injection. These approaches can reduce the knock and backfire problems, but they also reduce the specific power developed by the engine. The reduction is significant, as much as 40 to 50 percent, and would result in either a vehicle with reduced performance with the same size engine or a vehicle with the same performance with a much larger engine. Both of these effects are undesirable.

Research sponsored by the U.S. Department of Energy at the University of Miami, Florida, has resulted in laboratory demonstration of a fuel injection scheme that could potentially lead to a desirable method of induction of the fuel into the engine without either backfire or specific power reduction penalties.[7] It was found that, if the hydrogen was injected directly into the cylinder just after intake valve closure, a low-pressure injector could be used and no backfire or knock occurred. In this mode, the specific power was actually enhanced. This injection system would require further engineering improvement before it would be suitable for installation on a standard automobile, but the problems appear to be of the type easily solved by engineering development.

There is thus only one real problem requiring significant engineering effort in converting a vehicle to hydrogen--the technology of the fuel tank. Fuel tank criteria include reasonable volume and weight, a refuel time of no more than 10 minutes, and adequate safety. The four options of metallic hydrides, high-pressure gas, chemical fuels synthesized from hydrogen, and cryogenic liquid were selected for small vehicle fuel tanks.

To evaluate these options for the hydrogen fuel tank, a reference automobile with the following characteristics was selected as the baseline: weight, 1800 kilograms; range, 500 kilometers; fuel capacity, 80 liters (55.3 kilograms) gasoline; and mileage, 6.25 kilometers per liter.

These characteristics establish a baseline amount of fuel required to operate an automobile of specific performance, weight, and range. Ordinary gasoline is of variable composition and has a heat of combustion of 11 to 14 kilowatt hours per kilogram. The reference gasoline was assumed to be pure iso-octane, with a heat of combustion of 12.76 kilowatt hours per kilogram. The fuel capacity of the reference automobile is then equivalent to 706 kilowatt hours of energy. Hydrogen has a heat of combustion of 33.6 kilowatt hours per kilogram, so it would take 21 kilograms of hydrogen energy to equal the energy in a tank of gasoline. This amount was used to estimate the weight and volume of the hydrogen fuel tank system.

Metallic hydrides have attracted interest as carriers of hydrogen because they can be stored at ambient temperature and at ordinary pressure. This would seemingly lead to a simple storage tank manufactured in a manner much like today's gasoline tank. Unfortunately, the simplicity is an illusion, and factors such as extracting or releasing the hydrogen, safety, weight, volume, etc., will make application of the hydrides difficult.

The data in Table 6-1 were prepared in order to evaluate the hydride storage methods for hydrogen. The properties of the hydrides suggested as hydrogen storage media are provided, without consideration for the weights and volumes of the tankage, special equipment, or reactants necessary to extract the hydrogen from the hydride. The table is complete for all elements through titanium, with the exception of the rare gases. As can be seen, the weight percentage of hydrogen decreases as the atomic weight of the element increases. The table was terminated at titanium because it was felt that the heavier element hydrides contained so little hydrogen on a weight basis that they would be useless for automotive applications. Several alloys of the heavier

Table 6-1. Potential Carriers of Hydrogen

Atomic No	Name	Formula	Percent hydrogen	Weight required to provide 21.0 kg hydrogen	Specific gravity	Volume(L) required to provide 21.0 kg hydrogen	Safety problems	Ignition temperature	Performance Gates (+ = Pass, - = Fail) 1	2	3	4	5	6	Overall Pass
Simple hydrides															
1	Hydrogen, Liquid	H_2	100	21.0	0.07	300	Cold −252°C, fire	585°C	+	+	+	+	+	+	+
3	Lithium Hydride	LiH	12.68	166.0	0.82	202	Caustic, extremely hot fire on combustion	High	+	+	+	+	+	-	-
4	Beryllium Hydride	BeH_2	18.28	115.0	0.6?	191	Extremely hot fire, extreme toxicity	Low	+	+	+	+	-	-	-
5	Diborane	B_2H_6	21.86	96.0	0.417	214		Very low	+	+	+	+	+	-	-
6	Methane	CH_4	25.13	83.5	0.415	201	Toxic Cold −175°C, fire	575°C	+	+	+	+	+	+	+
7	Ammonia	NH_3	17.76	118.0	0.817	145	Toxic 100 ppm max, cold −30°C	High	-	+	+	+	+	-	-
8	Water	H_2O	11.19	188.0	1.000	188	–		-	-	-	-	-	-	-
9	Hydrogen Fluoride	HF	5.04	416.0	1.005	414			-	-	-	-	-	-	-
11	Sodium Hydride	NaH	4.20	500.0	0.92	543	Caustic, extremely hot fire on combustion	High	+	+	+	-	-	-	-
12	Magnesium Hydride	MgH_2	7.66	275.0	1.45	190	Extremely hot fire on combustion	Very low	+	-	-	-	-	-	-
13	Aluminium Hydride	AlH_3	10.08	208.0	1.3?	160?	Extremely hot fire on combustion	Very low	+	-	+	+	+	-	-
14	Silane	SiH_4	12.55	167.0	0.68	247	Toxic 0.1 ppm	−100°C	+	-	-	-	-	-	-
15	Phosphene	PH_3	8.89	236.0	0.746	316	Toxic 0.01 ppm, extremely hot fire on combustion	+20°C	+	-	-	-	-	-	-
16	Hydrogen Sulphide	H_2S	5.92	354.0	1.347	263	Toxic 20 ppm toxic combustion products	Low	+	+	-	-	-	-	-
17	Hydrogen Chloride	HCl	2.77	758.0	1.256	606			+	-	-	-	-	-	-
19	Potassium Hydride	KH	2.51	837.0	1.47	569	Caustic, extremely hot fire on combustion	High	+	-	-	+	+	+	-
20	Calcium Hydride	CaH_2	4.79	438.0	1.9	230	Extremely hot fire on combustion	Medium	+	-	+	+	+	+	-
21	Scandium Hydride	ScH_2	6.30	333.0	2.0?	170?	Toxic, extremely hot fire on combustion	?	+	-	+	+	+	-	-
22	Titanium Hydride	TiH_2	4.40	520.0	3.9	133	Extremely hot fire on combustion	High	+	-	+	+	+	-	-
Complex hydrides															
	Lithium Borohydride	$LiBH_4$	18.51	114.0	0.666	171	Mild toxicity	High	+	+	+	+	+	-	-
	Aluminium Borohydride	$Al(BH_4)_3$	16.91	124.0	0.545	228	Mild toxicity	Very low	+	-	+	+	-	-	-
	Magnesium Borohydride	$Mg(BH_4)_2$	14.93	141.0	1.3?	108?	Mild toxicity	Very low	+	-	+	+	-	-	-
	Beryllium Borohydride	$Be(BH_4)_2$	20.84	101.0	0.9?	113	Extreme toxicity	Low	+	+	+	+	-	-	-
	Lithium Aluminium Hydride	$LiAlH_4$	10.62	197.0	0.917	216	Toxic 10 ppm	Low	+	+	+	+	+	-	-
	Hydrazine	N_2H_4	12.58	167.0	1.011	163	Mild toxicity 500 ppm	300°C	+	+	+	+	+	+	+
	Iso-octane	C_8H_{18}	15.88	132.0	0.6918	191		270°C	+	+	+	+	+	+	+
Hydrogen absorbers															
	Palladium Hydride	Pd_2H	0.471	4458	10.78 cal	413		Low	+	+	-	-	-	-	-
	Lanthanum Nickel	$LaNi_5H_6$	1.38	1522	7.56 cal	201		Low	+	+	-	-	-	-	-
	Iron Titanium	$TiNiH_2$	1.87	1123	5.47	205		Medium	+	+	+	+	+	+	+

elements have been proposed as carriers for hydrogen on automobiles. These are included for historical interest, but their hydrogen content is so low that the weight penalty for their use would be prohibitive in automotive applications.

Several buses and automobiles have been equipped with metallic hydride storage systems by Billings Energy Research Corporation[8], Denver Research Institute[9] (a forklift) and, in Europe, Daimler Benz AG.[10] The systems use the iron titanium hydride enclosed in a tuble and shell heat exchanger for the removal of heat during charging and the addition of heat during discharging. They have a mass of over 500 kilograms and hold only enough hydrogen for about a 100- to 200-kilometer range. Both Billings and Daimler Benz use massive dilution with water or exhaust gas for backfire and knock control, but the great weight of the fuel tank strongly compromises the performance of the vehicle. The high dilution, with its accompanying loss of power, aggravates the performance loss due to the excessive weight of the hydride tank. The hydride systems also require 1/2 to 2 hours to refill, and the hydride heat exchanger must be attached to an external cooler during the charging cycle. These problems can be handled for a research demonstration vehicle but would be far too burdensome for a private automobile.

Some of the lightweight metallic hydrides have relatively high hydrogen content, but the release of the hydrogen requires the addition of a second reactive chemical or temperatures over 500°C. To evaluate these materials, the following specific performance gates were selected; the criteria for these gates were arbitrarily established with the hope that they would at least be reasonable:

1. The hydride should be a potential fuel.

2. There should not be a solid residue. (Recycling a solid residue would require a return logistics network as extensive as the supply network, which would result in a large cost penalty for using the material.)

3. The potential fuel to supply the required energy should not weigh more than four times the weight of gasoline (221.2 kilograms).

4. The potential fuel should not occupy more than four times the volume of gasoline (320 liters).

5. There should not be any unusual safety problems.

6.   The chemical elements used as carriers must be sufficiently abundant to support the conversion of a significant number of automobiles to their use.

The hydrides are compared and their ability to pass or fail the performance gates is shown in Table 6-1, with only hydrogen and the hydrocarbons we wish to replace surviving the evaluation process.

By ignoring toxicity, ammonia and hydrazine might be considered. Both of these materials give a warning of their presence by a strong odor, so the danger of accidental poisoning is low. However, in the case of accidents in which the fuel tank is ruptured, the hazard potential can be very great. One deep breath of either hydride would have a high probability of being fatal. Ammonia, along with liquid hydrogen, presents a potential for extensive frostbite if the tank is ruptured and fuel splashes on the passengers. Hydrazine, like gasoline, is very persistent; because it has a boiling point higher than water (114°C), any spilled material will be present in the area of the accident for a long time. Persons involved in the accident are therefore subjected to a much longer exposure to the toxicity and fire hazards. In addition, hydrazine is currently classified as a possible carcinogen.

While there may be some hesitation about these fuels because of their toxicity, the most telling problems are the manufacturing methods and the cost penalty their use incurs. Currently, the only synthetic route to ammonia production is the direct reaction of hydrogen and nitrogen, and the most economic method of producing hydrazine is from ammonia. The cost of producing these fuels will thus be quite high on the basis of the energy content, much higher than the cost of the starting product, hydrogen.

HIGH-PRESSURE GAS

High-pressure gas must be examined because it can be adapted for use with the least development effort. As a result, many of the hydrogen-fueled automobiles built so far have used this fuel tank technology.

The evaluation of high-pressure cylinders for storage can be approached from a standpoint similar to that used with the hydrides, and the same performance gates can be used. The hydrogen fuel charge reference amount is 21.0 kilograms. Containment of 21.0 kilograms of hydrogen in a 320-liter volume requires a pressure of 800 atmospheres, calculated from simple ideal gas laws. The actual pressure would be higher than 800 atmospheres because of high-pressure deviations from ideal behavior. To contain this pressure in steel with a tensile strength of 10^8 kg m^{-2}, a sphere would require a wall thickness of 7 centimeters and would weigh 1550 kilograms. Stronger materials of higher cost are available; for example, cryo-formed 304L stainless or maraging steel, titanium, or composite materials containing glass, boron, or carbon filaments. These materials could potentially be fabricated into pressure vessels that would contain hydrogen at pressures of over 800 atmospheres and meet the desired weight. However, little work has been performed on developing such containers for use in automobiles, and there is doubt about whether this type of container would be feasible.

If high-pressure tanks can be developed that would contain hydrogen at these high pressures, they would be quite easily adapted to the fuel system of the current internal combustion automobiles. But questions would still remain regarding the safety of high-pressure gas storage in the private automobile. If one of these tanks were crushed in a fully charged state, the pressure-driven expansion of the gas would be difficult to characterize as other than an explosion. Care in the design of the automobile would be necessary to protect the passengers from injury should a rupture occur. Pressure vessels made from composits of high-strength filaments and a binder have one great safety advantage over metal tanks: they do not form fragments when they burst. On puncture, a metal tank tends to break into pieces that are thrown about like shrapnel; a composit tank frays at the puncture but otherwise remains intact. Much work will be required to demonstrate the use of this technology and its safety before actual use could be contemplated.

CHEMICAL FUELS SYNTHESIZED FROM HYDROGEN

The large-scale use of hydrogen as a fuel would also make possible the synthesis of a number of liquid fuels that could be used directly in current technology engines. Although most of these have already been evaluated as hydrogen sources, some examination of their characteristics when used directly as fuels is appropriate. Table 6-2 compares these synthetic fuels on a simple weight and volume basis with the reference iso-octane. Compared to gasoline-type fuels, all the synthetic fuels require significantly more volume for storing a unit amount of energy. Although hydrogen is by far the least desirable from the volume standpoint, it is by far the best on a weight basis.

All the manufactured fuels should be more expensive than the hydrogen from which they will be synthesized. Ammonia and hydrazine, as previously mentioned, present real problems because of their toxicity, and the carbon-containing fuels present the same environmental problems being fought with the current fossil fuels. The

Table 6-2. Chemical Fuels
Synthesized from Hydrogen

		Volume (litres)	Weight (kilograms)	Volume ratio	Weight ratio	Energy (kilowatt hours/ kilogram)
	Iso-octane	80	55.7	1.00	1.00	12.76
	Ammonia	167	136.7	2.09	2.47	5.16
	Hydrazine	151	152.5	1.89	2.76	4.63
Fuels necessary to provide 706 kilowatt hours of energy, equal to 80 litres of gasoline	Hydrogen	300	21.0	3.75	0.38	33.60
	Methane	122	50.7	1.53	0.92	13.92
	Methanol	158	124.9	2.26	1.97	5.65

advantage offered by these fuels is only one of convenience. If liquid hydrogen or high-pressure gas can be made safe and convenient, this advantage will no longer be significant.

Cryogenic Liquid Hydrogen

Compared to the other methods of storing hydrogen, cryogenic storage has the following advantages and disadvantages:

- Advantages

    - Lowest cost per unit energy
    - Lowest weight per unit energy
    - Simple supply logistics
    - Normal refuel time
    - Probably as low an implementation cost as any technology except for high-pressure gas
    - No insurmountable safety problems

● Disadvantages

    - Loss of fuel when vehicle is not in use
    - Large tank size
    - Cryogenic liquid safety engineering problems

There is thus a strong argument supporting liquid hydrogen as the only practical form of hydrogen for vehicular transport as an alternative to the hydrocarbon fuels. The use of liquid hydrogen will produce an enormous growth in the cryogenic industry. To realize this growth, several areas of technology will have to be advanced, and some new hardware developed. These areas can be outlined as follows.

The bulk transport of commercial hydrogen across the continent will be by pipeline. Two options exist for the liquefaction of this pipeline-supplied gas. Each vehicle refueling station could have a small liquefier and storage tank to supply its own needs. This option would require the development of small, highly reliable liquefiers. Regional liquefaction plants with large-capacity units and large storage facilities from which liquid hydrogen could be distributed by truck are the second option. The regional approach would require liquefaction and storage facilities significantly larger than any currently in use. For the United States alone, the total magnitude of this supply would be on the order of 10^{13} liters per year. If this were divided evenly between 100 regional liquefaction centers, the production rate of each would be 10^{11} liters per year, or 3×10^8 liters per day. These production rates are much higher than that of any existing liquid hydrogen production facility.

Transfer of the liquid hydrogen at various points in the supply system will present some specific development problems. For the regional liquefaction option, transfer will be required from large stationary storage containers to over-the-road delivery trucks and from the trucks to the local retail service point. Current hardware and trucks are probably suitable and usable. For either the regional or local liquefaction option, much development is required to provide the simple, fail-safe, fool-proof hardware necessary to refuel the private automobiles. It may be desirable and feasible to develop totally automatic refueling hardware and systems.[11]

Storage of the liquid hydrogen on the vehicle will require the development of highly reliable, low heat leak, low-cost storage vessels. These vessels will be similar in design to current small liquid hydrogen storage vessels--double-walled, vacuum-jacketed, etc.--but will be modified to fit the envelope of the automobile and to meet the necessary safety requirements for mobile use. Some work has already been performed in this area in adapting automobiles to run on cryogenic liquid natural gas.

Heat exchangers will be required to warm the cold gas or liquid to ambient temperature. If gas is withdrawn, it will be necessary to provide tank pressurization by heating the liquid. If liquid is withdrawn, heat or helium can be used to provide the pressure. The liquid withdrawal mode enables the cooling capacity of the hydrogen to be used to provide air conditioning for the vehicle. The obvious source of heat for heating the hydrogen to ambient temperature is the engine exhaust gas. Using this gas will require care to prevent frost buildup or similar problems from the condensation of water.

Changes to the engine are probably minor. As long ago as 1948, it was demonstrated that a carburetor engine free from carbon deposits will run without problems at compression ratios as high as 10 to 1 using hydrogen as the fuel. Billings Energy Research Corporation has converted several automobiles to hydrogen fuel and, in 1976, a bus whose route was between Provo and Orem, Utah. Currently, Professor Robert Adt of the University of Miami is studying the detailed changes necessary to convert ordinary engines to hydrogen fuel use. His work is being supported by the Department of Energy under the direction of Eugene E. Ecklund, and results should be available to automobile designers in 1981.[7]

Safety devices will be required to prevent hazards from the use of hydrogen fuel. In a well-designed system that is used as often as every other day, venting will not be required. Most vehicles are used at least this often, but some will not be, and every vehicle is occasionally allowed to stand unused for several days. It is therefore necessary to have a safe venting system for these vehicles. Because of the high dispersal characteristics of hydrogen due to its buoyancy and rapid diffusion, it may be safe to simply vent it slowly into the ambient air as is done with gasoline. However, extensive safety testing would be required to demonstrate this to everyone's satisfaction. An alternative to positive safe venting could be obtained by catalytic ignition, electric ignition, or resonance tube ignition, all of which would require the design of a small flameless burner and stack for product removal.

Catalytic ignition will require development. Currently, catalysts can be made from platinum and palladium that will spontaneously ignite hydrogen. They will work well but will be quite expensive. Raney nickel can also be used as a catalyst. It is much less costly but is very easily poisoned. A new catalyst material based on lanthanum cobalt oxide ($LaCoO_3$) and similar compounds may perform this function well and be quite inexpensive.

Electric ignition would be quite inexpensive but would deplete the battery at a time when no recharging would be available. Over a period of time, the battery could become sufficiently depleted that it would no longer start the engine. An intriguing possibility for venting would be through a small hydrogen/air fuel cell to provide a small amount of electric power. The power could be used to charge the primary battery during periods when the car was not in use. Implementation of this scheme would eliminate low-battery-induced starting failures.

A resonance tube igniter might be quite reliable and inexpensive, but would require high-velocity gas to function. To obtain the necessary high velocity, the vent gas would have to be at a pressure of at least 10 to 20 atmospheres, which would not be compatible with a low-pressure liquid storage system.

The final consideration is protection against collisions of sufficient magnitue to rupture the fuel tank. By the very nature of its design, the liquid hydrogen fuel tank will be more difficult to rupture than a conventional gasoline tank and, when ruptured, the hydrogen would dissipate more rapidly than gasoline. The primary safety consideration is therefore to locate the tank in a position to minimize its involvement in the statistically most common frontal collision and to protect the passenger compartment from the splashing liquid by lightweight splash screens and such.

The use of hydrogen in internal combustion engines would nearly eliminate their production of air pollution without lowering the high standards of performance or range to which we have become accustomed. Although on a weight basis hydrogen is a much better fuel than gasoline, it is much less dense, so a given weight takes up more space. These two effects work together to produce a fuel tank for a 300- to 400-mile range vehicle about three times as big and half as heavy as the current fuel tank on the average automobile. Although undesirable, the increase in bulk could easily be accommodated within the size limitations of current automobiles with a tolerable reduction in available trunk space, as shown in Figure 6-4. Figures 6-5 and 6-6 show the schematic layout of the liquid hydrogen system within the automobile and the service station.

Conventional internal combustion engines burning hydrogen would still produce trace amounts of pollution as the result of oil burning. In new engines this pollution would be minimal, but as engines become old and worn, more oil would be burned. This source of pollution could be reduced by some motor design changes and by altering the composition of the oils used for lubrication. The absence of carbon, lead deposits, and corrosive combustion products in hydrogen-burning engines would greatly decrease the wear rate of the engine and thus reduce the rate at which oil burning increases as an engine becomes older.

Conversion of the automobile to hydrogen hardware development required

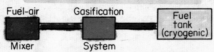

Fuel-air   **Gasification**   **Fuel tank (cryogenic)**

Mixer   **System**

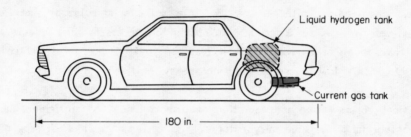

Liquid hydrogen tank

Current gas tank

180 in.

	Gasoline	Hydrogen
Range	650 km (400 miles)	650 km (400 miles)
Tank size	80 liters (21 gallons)	312 liters (83 gallons)
Fuel weight	56 kilograms (123 lbs)	21.8 kilograms (48 lbs)
Fuel consumption rate	{ 8.1 kilometers/liter	2.1 kilometers/liter
	{ 11.6 kilometers/kilogram	30 kilometers/kilogram

For trucks and buses the components will be larger

Pressurization heater

Fill and use line

Vacuum jacket

Vehicle exterior

Fill and drain port

60 cm

Tank

130 cm

Ambient temperature gas

Warmed gas

Air

Air-hydrogen mixer, volumetric

Engine

Heat exchanger

Exhaust gas

Output

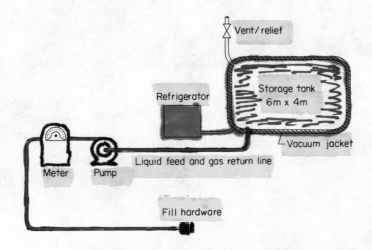

Schematic Layout of a Liquid Hydrogen Cryogenic
Service Station for Liquid-Hydrogen-Fueled
Automobiles.

A number of automobiles have been converted to operate on hydrogen, such as
the standard American automobile shown in Figure 6-7, which uses an experimental
metallic hydride tank for storage and a conventional propane type carburetor for
premixing the hydrogen and the air. Figures 6-8 and 6-9 show the engine conversion
and the fuel tank. To prevent backfires, it was necessary to use water injection and
exhaust gas recirculation in this conversion, which resulted in a dramatic reduction in
the power developed by the engine. The hydride tank weighed nearly 500 kilograms and
contained only enough fuel for a few tens of kilometers of operation. Charging the
tank required the attachment of a secondary coolant loop outside the automobile to
remove the heat of formation of the metallic hydride; recharging required several
hours. This vehicle clearly demonstrated that it was possible to operate a relatively
conventional automobile with hydrogen as a fuel, but its lack of performance and short
range revealed that much more engineering work would be required to produce a vehicle
that would satisfy the needs and desires of the private automobile owner.

In 1979, the Denver Research Institute, Denver, Colorado, converted a small
truck with a stratified charge engine and a supercharger to operate on hydrogen. Preliminary results indicate that this vehicle has no backfire problem and no reduction
of engine power. In the 1979 configuration, as shown in Figure 6-10, no specific
vehicle fuel tank had been designed for the vehicle and it was operated on conventional
steel high-pressure gas bottles. Further testing will be required to determine if the
technology demonstrated in this vehicle will be suitable for wider use.

Figure 6-7. A 1975 Pontiac Grand Ville Converted to Operate
            on Hydrogen (Courtesy of the International
            Association for Hydrogen Energy, P.O. Box
            248294, Coral Gables, Florida 33124, U.S.A.).

Figure 6-8. The Carburetor Area of a Conventional
            450-Cubic-Inch Engine that Billings Energy
            Research Corporation has Converted for Operation
            on Hydrogen (Courtesy of the International
            Association for Hydrogen Energy, P.O. Box 248294,
            Coral Gables, Florida 33124, U.S.A.).

Figure 6-9.   Rear View of 1975 Pontiac Grand Ville, Showing
a Metal Hydride Tank, Which Has Been Mounted in
the Trunk for Display Purposes and Which Replaces
the Conventional Fuel Tank. (The metal hydride
in a dry powdery form appears to offer one possible
method of storing hydrogen for vehicular use.)
(Courtesy of the International Association for
Hydrogen Energy, P.O. Box 248294, Coral Gables,
Florida 33124, U.S.A.).

Figure 6-10. A Dodge D-50 Truck Converted to Hydrogen by the Denver Research Institute, Denver, Colorado, for the Clean Fuels Institute, Riverside, California (Courtesy of the Clean Fuels Institute, Riverside, California, U.S.A.).

With proper design, conversion of an existing automobile to hydrogen use or construction of new units using this fuel would lead to a machine that looks, sounds, and performs essentially as automobiles do now. Existing fueling stations would still be useful for refuel and repair. Thus, except for the elimination of automotive fuel pollution, the public would not be subjected to any drastic changes in automobile use.

HYDROGEN-FUELED AIRCRAFT

For application to airplanes, the problem resolves itself entirely to physical bulk. On a weight basis, hydrogen produces 2.8 times as much energy as hydrocarbon fuels (33.60 kilowatt hours per kilogram for hydrogen versus 12.0 kilowatt hours per kilogram for hydrocarbon fuels), but liquid hydrogen has only about 0.1 the density of hydrocarbons (0.07 kilogram per liter for liquid hydrogen versus 0.7 kilogram per liter for hydrocarbons). This converts to an enegy ratio on a volume basis of 0.35 times as much energy per liter of fuel space using hdyrogen as using hydrocarbons. Two other factors enter into consideration: First, a significant weight of insulation is required to thermally protect the liquid hydrogen (a negative effect); second, the propulsive efficiency of a hydrogen-burning jet engine is significantly higher than an engine burning hdyrocarbons, because of the lower molecular weight of the exhaust products (a positive effect). These factors, working together, result in a significant improvement in performance for hydrogen-fueled airplanes that would result in a range increase of 1.5 to 2.5 times that of a plane of the same weight that used hydrocarbon fuel.[12]

It will be perfectly feasible and, in fact, advantageous to design all new aircraft to use hydrogen as the fuel. It should also be possible to convert most existing large aircraft to hydrogen use by adding wing tanks for hydrogen storage. Considerable experimentation has been done on the combustion of hydrogen in gas turbine engines, and this experience indicates that it is quite simple to make the conversion and that reliable, trouble-free operation is obtained.

Both Lockheed and Boeing aircraft corporations have performed feasibility studies of subsonic and supersonic aircraft fueled by hydrogen. The studies have all indicated that there are no insurmountable technical barriers to the use of hydrogen fuel. In 1957, under the sponsorship of the National Aeronautics and Space Administration (NASA) Lewis Flight Propulsion Laboratory (currently the NASA Lewis Research Center), a B-57 "Canberra" bomber was equipped so that one engine could be converted to hydrogen fuel during flight. This aircraft was tested with both pressure-fed and pump-fed liquid hydrogen fuel supplied to the engine. Hydrogen fueled flights of as long as 20 minutes were achieved, and the program met all the expected goals.[13] During the same time period, Pratt & Whitney developed the 304 "hydrogen expander" engine, which was designed to use the unique properties of liquid hydrogen fuel.[5] This engine was also quite successful but, at that time, the difficulty of obtaining liquid hydrogen and the logistics problems of supply to the aircraft caused termination of the program. (This happened before the Apollo program developed the capacity to fly the liquid-hydrogen-powered SIB Saturn 5 second stage.) No technical problems were discovered that would preclude the future use of hydrogen as a fuel for aircraft.

Recently, Daniel Brewer of Lockheed, California, partly under the sponsorship of Robert Witcowski of NASA Langley, performed several definitive studies of the feasibility of a commercial subsonic passenger-carrying aircraft.[14] A number of configurations of the aircraft were examined for these studies, dealing with the placement in the airframe structure of the rather bulky hydrogen fuel tanks. All but two were rejected in the initial screening because no advantage could be found for exotic airframe designs over the current layout. The two relatively conventional designs that were selected for study are shown in Figure 6-11 . These two configurations were subjected to detailed study and compared to aircraft with identical capabilities fueled with the currently used jet-A fuel. The configuration with the internal tanks was found to be either equal to or superior to the exterior wing tank configuration in almost all respects. The internal tank configuration was then compared to a similar performing jet-A fueled aircraft. Table 6-3 indicates the results of this comparison. The hydrogen-fueled aircraft was found to be superior to the jet-A fueled aircraft in all respects except that it is a little longer (because of the space required to store the hydrogen) and will need a slightly longer runway to land. Unmentioned in the table are the advantages of the very low air pollution and significantly quieter operation of the hydrogen aircraft, quieter even than the quietest large aircraft now in service. Thus, there are compelling technical reasons to design the next generation of subsonic passenger aircraft to use hydrogen as the fuel. The barriers are in the area of liquid hydrogen supply and the cost per unit of energy.

Lockheed, California, has been attempting to interest a multinational group in funding a demonstration of a liquid-hydrogen-fueled subsonic aircraft. At a meeting in Stuttgart, Germany, in September 1979, the United States, Canada, Britain, France, Saudi Arabia, Switzerland, and Japan agreed that a joint program, managed by an international energy agency, should be launched to demonstrate the feasibility of liquid hydrogen as an aircraft fuel. The plan is to convert several existing L-1011 aircraft to hydrogen for freighter service. The conversion will require the stretching of the fuselage enough to accommodate the liquid hydrogen tanks. The cargo area will be about the same size as is currently available in the L-1011. Two hydrogen tanks will be required to allow for balance of the aircraft as the fuel is removed, and the cargo

Figure 6-11. Conceptual Designs for Liquid-Hydrogen-Fueled
            Aircraft Investigated by Lockheed-California
            Company Under Contract to NASA, Langley Flight
            Research Center. (The upper aircraft has
            external wing-mounted liquid hydrogen tanks; the
            lower has the tank mounted internally.)
            (Courtesy of Lockheed-California Company, Burbank,
            California 91520, U.S.A.).

area will be between the two tanks. Four airports in four different countries will be
equipped to service and fuel the converted aircraft, which will fly a circuit through
these airports carrying freight and demonstrating the performance of the liquid-
hydrogen-fueled airplane. An artist's conception of the appearance of the converted
L-1011 is shown in Figure 6-12.

The high energy requirements for supersonic flight make hydrogen the natural
choice and, because of its high energy content, perhaps the only practical fuel for
these aircraft. Although none has yet flown, Figures 6-13 and 6-14 show an artist's
conception of two such supersonic hydrogen-fueled aircraft.

Table 6-3.   Comparison: Jet A VS LH_2 Passenger Aircraft
(5500 n mi, M 0.85)

		LH_2	Jet A	Factor (Jet A/LH_2)
Gross weight	lb	391,700	523,200	1.34
Operating empty weight	lb	242,100	244,400	1.01
Block fuel weight	lb	52,900	165,500	3.13
L/D (cruise)		16.1	17.9	1.11
Thrust per engine	lb	28,700	32,700	1.14
Span	ft	174	194.1	1.12
Height	ft	59.5	60.2	1.01
Fuselage length	ft	219	197	0.90
Wing area	ft^2	3363	4186	1.24
FAR T.O. distance	ft	6240	7990	1.28
FAR landing distance	ft	5810	5210	0.90
Aircraft price	$$10^6$	26.9	26.5	0.99
Energy utilization	Btu/seat n mi.	1239	1384	1.12

Courtesy Lockheed-California Company, Burbank, California 91520, U.S.A.

HYDROGEN FUEL CELL

One of the great benefits of the adoption of hydrogen as the general-purpose fuel will be the ability to use the hydrogen-air fuel cell. Hydrogen-air fuel cells are, in many respects, the reverse of the electrolysis cell for the production of hydrogen from water. These cells accept hydrogen from storage and oxygen from the air and produce electric power and water. Like the electrolysis cell, they are quite efficient. When in operation they are nearly silent, with only the whisper of flowing gas to indicate they are operating. They can be manufactured in sizes ranging from small enough to power a flashlight to large enough to power a small town.

Early models of fuel cells were used to provide the operating power on the Gemini and Apollo spacecraft. These fuel cells required pure oxygen for their operation rather than oxygen taken from the air, but the principle of operation of the hydrogen-oxygen fuel cell and the hydrogen-air fuel cell is the same. The cells used in the manned spaced missions operated with very high reliability, providing power for all spacecraft functions. The water co-produced with the electric energy was pure distilled water and was used without treatment as the drinking water for the crew of the space-craft. These early fuel cells were rather expensive to build, partly because they were designed and built as near to perfection as possible to ensure the safety of the astro-

Figure 6-12. An Artist's Conception of a Lockheed Liquid-Hydrogen-Fueled L-1011 Freighter. This Plane could Become a Reality by 1985 (Courtesy of Lockheed-California, Burbank, California 91520, U.S.A.).

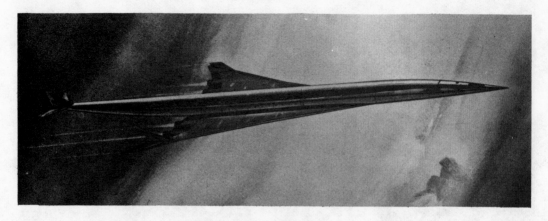

Figure 6-13. An Artist's Conception of a Mach 2.7
(3200 kilometers/hour) Liquid-Hydrogen-Fueled
Aircraft Carrying 234 Passengers, 6800 Kilometers
(Courtesy of Lockheed-California Company, Burbank,
California 91520, U.S.A.).

Figure 6-14. An Artist's Conception of a Mach 6
(7200 kilometers/hour) Liquid-Hydrogen-Fueled
Aircraft Carrying 200 Passengers, 8000 Kilometers
(Courtesy of Lockheed-California Company, Burbank,
California 91520, U.S.A.).

nauts, but partly because they used expensive platinum catalysts on the electrodes. In the years since the original spacecraft fuel cells were designed and built, much progress has been made toward the construction of fuel cells of less scarce and costly materials.

Like the electrolyzers with which they have so much in common, the fuel cell has two electrodes separated by an electrolyte. The electrolyte can be a basic solution of sodium or potassium hydroxide; an acidic solution, perhaps of phosphoric acid; or one of the various solid ceramic or polymeric solids that carry electric current in the form of hydroxyl ions or hydrogen ions. At the anode, hydrogen gas reacts to produce hydrogen ions, H^+, and electrons. The electrons are driven through the external circuit, where they can do work, and finally to the cathode, where they react with the oxygen and the electrolyte to produce hydroxyl ions, OH^-. In the electrolyte, the hydroxyl ions and the hydrogen ions react to produce water. The water escapes from the cell. The nature of the chemical reactions that occur are essentially the same as those that occur in a battery, but the physical process is quite different. It is to this difference that the fuel cell owes its great advantage. Figure 6-15 diagrams the essential features of the fuel cell.

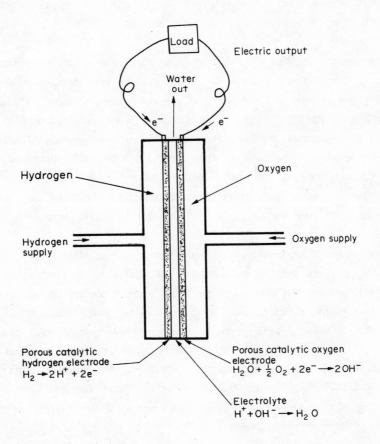

Figure 6-15. Schematic Diagram of a Hydrogen-Oxygen Fuel Cell for the Direct Conversion of Hydrogen to Electric Power.

In a battery, the substance that gives up electrons and is chemically oxidized and the substance that receives the electrons and is chemically reduced are both present at all times. In a common zinc-carbon maganese dioxide battery, the zinc is oxidized to zinc chloride, the manganese dioxide is reduced to manganous oxide, and water is produced. All remain in the cell when it is discharged, such that the weight of the battery is the same when fully charged as when fully discharged. The same set of circumstances is true for any of the several types of rechargeable batteries. Even though the chemistry is the same, the mass balance for the fuel cell is enormously more favorable. The structure of the fuel cell only directs the reactions and does not itself take part in them. The hydrogen is the only consumable commodity that must be carried. The hydrogen enters the fuel cell, where it reacts with oxygen that was extracted from the air; these in turn react to form water that is rejected to the air and carried away as vapor. The hydrogen-air fuel cell system loses the weight of the hydrogen as it is consumed. The weight of the second reactant need not be accounted for because it is extracted from the air as it is used.

It is necessary to digress for a moment to place a proper perspective on the severe penalty that battery systems for vehicles must bear because of the requirement to carry both the oxidant and the fuel. If the reference automobile discussed earlier in this chapter, with its 55.6 kilograms of gasoline, were also required to carry the air necessary to combust this gasoline, it would need a tank capable of holding somewhat over 1025 kilograms of air and another tank to hold 1080 kilograms of combustion products. Clearly, one of the major reasons that the gasoline-powered vehicle is so succesful is that it takes one of its chemical reactants from the air and discards the waste products back into the air. The hydrogen-air fuel cell shares this advantage.

Fuel cells also offer greatly improved thermodynamic efficiency. The hydrogen-air fuel cells can convert the potential energy of hydrogen to electric energy at about 80-percent efficiency. An electric motor can convert electric power into shaft power at about 95-percent efficiency. Combining these two technologies into a drive mechanism for a vehicle would provide a system that would have about 75-percent efficiency. The fuel cell is somewhat more efficient at low levels of power output than at high levels, and electric motors have about the same efficiency at all rotational speeds within their design characteristics. These factors combine to provide a potential design for a vehicle that would perform at about 70-percent thermodynamic efficiency over all operational regimes.

In a laboratory test environment, a carburetted internal combustion engine can be operated at efficiencies of 35 percent, and a direct injection diesel at about 45 percent. To obtain this efficiency, the operation speed of the engine must be adjusted to the optimum for the engine, usually a speed somewhat slower than the speed for maximum power. At both lower and higher speeds, the efficiency is reduced. In an actual operational environment, the average efficiency delivered is in the range of 20 percent for the carburetted engine and 25 percent for the diesel. The hydrogen-air fuel cell system is therefore 3.5 times more efficient than the carburetted engine and 2.8 times more efficient than the diesel. In practice, this improvement in efficiency would result in several very important effects. Given a fuel cell vehicle with the same range and performance as the reference automobile, the hydrogen fuel tank would have to hold only one-third as much hydrogen as it would if the hydrogen were burned in an internal combustion engine. To maintain the same cost per mile as the reference automobile, the hydrogen could cost three times as much as hydrocarbon fuel, but the total energy consumed in performing the vehicle's task would only be one-third as great.

Development of the hydrogen-air fuel cell vehicle would be very advantageous. Such a vehicle would be very quiet, would emit no pollutants (fuel cells do not even emit the nitrogen oxides that are the only pollutant when hydrogen is burned in an internal combustion engine), and should be reasonably economic to operate. The adoption of the fuel cell would also alleviate part of the problem associated with the large bulk of low-density hydrogen necessary to achieve standard range when the hydrogen is used in an internal combustion engine. From Table 6-2, the volume of liquid hydrogen necessary for the reference automobile is 300 liters. Using a fuel cell, only 100 liters of fuel are required, or only 20 percent more than the volume of gasoline carried on the reference vehicle (80 liters). Development of this technology may make feasible the most easily implemented storage method, compressed gas. An advanced technology gas bottle made from high-strength filament materials such as glass fiber, carbon fiber, boron fiber, or the Dupont fiber Kevlar® could be developed that would be quite light. The three-fold reduction in the quantity of hydrogen required for the fuel cell vehicle would reduce the tank size such that the compressed gas tank would be acceptable.

Because of the great size flexibility of the fuel cell, this technology could be used for the whole spectrum of surface vehicles. From the lightest mo-ped to very large earth-moving equipment, these vehicles would be quiet, efficient, and essentially nonpolluting. The infrastructure necessary for the refueling of these vehicles would be much the same as that used for refueling hydrocarbon-powered vehicles. Some type of roadside service station would be used offering basically the same type of refueling service as the present gasoline stations. The station would receive its liquid hydrogen supply by over-the-road tanker in much the same manner that hydrocarbons are delivered today. If high-pressure gas were used for the vehicle fuel supply, the station could receive its supply by a modest pressure pipeline distribution system. A fuel cell within the station would use some of the piped hydrogen to provide electric power to operate a high-pressure pump, which would compress the hydrogen to the refueling pressure necessary for transfer to the vehicle.

The fuel could also be adapted to provide the bulk of the electric power currently transmitted by power lines. For this system, each user of electric energy, whether single home or large commercial facility, would install enough fuel cells to provide the electric power required and would use pipeline supplies of hydrogen to generate the power. In a study performed in 1968, it was found that the most economic system of energy distribution is the pipeline distribution of natural gas, with catalytic reformers at each consuming site to convert the natural gas to hydrogen and carbon monoxide, followed by the generation of electric power by fuel cell conversion of the hydrogen. This system was the least costly because of the inherent efficiency of the fuel cell and the very low cost of transmission of energy as a gas in a pipeline. Several other important advantages accrue from the use of this system, on which it is difficult to place a price. Experience in most parts of the world indicates that pipelines are extremely reliable. Most of us have experienced an electric power outage, but gas outages are very rare. Some of this reliability stems from the storge capability of a pipeline. A pipeline several hundred kilometers long has a large amount of gas stored within. If an interruption occurs somewhere along the line, the stored pressurized gas can be isolated from the rest of the pipeline and customers can be supplied for a period of time, during which the reason for the failure can be corrected. The customer is never aware that there was any trouble on the supply line. Gas pipelines can easily be buried under ground. This removes them from view and isolates them from the gross effects of weather changes. In addition, it enormously reduced the right-of-way needed. Underground electric cables are possible but are very expensive.

Although electrical energy can be transmitted, there are many problems. Electrical transmisssion lines are not perfect--some power is lost--and the longer the line, the more power converted to heat by the resistance of the lines. Distance is the key. or short distances, electrical transmission is quite good but, as the distances increase, power loss, cost, and technical difficulty also increase. When 60-hertz alternating current lines become an appreciable part of a wavelength long, a new loss factor comes into play, namely radiation. The critical distance for radiation loss at 60 hertz is 775 miles. A simple transmission line of this length would radiate a large part of the power put into it. This loss can be observed as a hum in the car radio when near high-tension transmission lines.

Two things can be done to reduce these losses: first, the simple resistance loss can be reduced by using the highest voltage possible. These very high voltages (up to

750,000 volts) require transmission towers that are very tall, and the transmission lines must be protected from accidental contact with anything because voltages this high can cause sparks several feet long.   The electrical transmission engineers have done a remarkable job of designing these systems to provide safety and reliability, but they still leave much to be desired.   Such lines have caused fires and other damage and do fail more often than desirable.   In addition, the corridors of clearcut forest necessary for the transmission lines, plus the appearance of the towers, are not very scenic.

The problem with radiation losses can be suppressed by the use of direct-current transmission lines.   These lines also require very high voltages to reduce resistance losses, and in addition require expensive complex power converters at each end of the line.   These converters take the very high voltage alternating current and convert it to very high voltage direct current at the input end of the line and perform the opposite conversion at the output end.   These convertors must use high voltage alternating current because there are no devices available equivalent to transformers for converting direct current voltages from high to low voltage and back again.   Even though there are electrical transmission lines in various places in the world that transmit energy over distances of a thousand miles, gaseous hydrogen is probably a better way to move energy over distances greater than 300 miles.

Electrical energy must be used as it is made; there is no storage in the transmission line.   Even the engineering marvel presented by high-voltage transmission lines is exceeded by the complex control networks required to match, on a minute-to-minute basis, the output of the various generators in a power grid to the consumption rate. The blackout of the U.S. East Coast in the 1960s was the result of an unpredicted change in the use rate of sufficient magnitude that the control network was unable to handle it.   The surge of unused power caused many other parts of the network to shut themselves off in an effort to protect themselves, with an increasing mismanagement of the load until the whole system failed.   Not only was there hardship caused by the outage, but many millions of dollars of damage was done to the equipment involved. The fact that these occurrences are not more common is a great tribute to the engineering skill of the systems designers and the dedication of the operators, not to the inherent reliability of this type of energy distribution system.

The use of hydrogen as a general-purpose fuel is not dependent on the development of the fuel cell.   However, when the hydrogen power system comes into use, a natural evolution to fuel cell use will follow.   The growth to the air-hydrogen fuel cell for transportation power will provide a system that is quiet, reliable, much more energy efficient, and basically nonpolluting.   The use of the fuel cell for the production of home and industrial electric power will provide a system that is more reliable and efficient and will allow the elimination of cross-country transmission of electric energy, with its high-cost, unsightly, and somewhat hazardous transmission towers.

Chapter 6 References

1. L.O. Barthold and H.G. Pfeiffer, "High Voltage Power Transmission," Scientific American, Vol. 210, No. 5.

2. N.P. Chopey, "Industry Joins Liquid-Hydrogen Scene," Chemical Engineering, Vol. 67, 1960.

3. C.F. Sindt, P. R. Ludtke, and D.E. Daney, "Slush Hydrogen and Fluid Characterization and Instrumentation," Tech Note 377, U.S. National Bureau of Standards, 1969.

4. E.J. Wortberg and F.R. Block, "Production of Iron with Electrolytic Hydrogen Through the Application of Nuclear Energy," Ver. Dtsch. Ing., Vol. 115, No. 16, November 1973.

5. R.C. Mulready, "Liquid Hydrogen Engines," Technology and the Uses of Liquid Hydrogen, Pergamon Press, McMillan Company, New York and London, 1964.

6. R.O. King and M. Rand, "Oxidation, Detonation, Ignition and Detonation of Fuel Vapors and Gases, xxvii Hydrogen Engine," Canadian Journal of Technology, Vol. 33, 1955.

7. R.R. Adt and M.R. Swain, "Hydrogen Engine Configurations and Design Approaches," Clean Energy Research Institute, University of Miami, 1979, in preparation.

8. R.L. Woolley, "Design Considerations for the Riverside Hydrogen Bus," Proceedings of the 2nd World Hydrogen Energy Conference, Zurich, 1978, Pergamon Press.

9. F.E. Lynch and E. Snape, "Technical and Economic Aspects of In-Plant Hydride-Fueled Fleet Vehicles," Proceedings of the 2nd World Hydrogen Energy Conference, Zurich, 1978, Pergamon Press.

10.   H. Buchner and H. Saufferer, "The Hydrogen/Hydride Energy Concept," Pro-
      ceedings of the 2nd World Hydrogen Energy Conference, Zurich, 1978, Pergamon
      Press.

11.   W. Peschka and C. Carpetis, "Cryogenic Hydrogen Storage and Refueling for Auto-
      mobiles," Proceedings of the 2nd World Hydrogen Energy Conference, Zurich,
      1978, Pergamon Press.

12.   W.J.D. Escher and G.D. Brewer, "Hydrogen: Make-Sense Fuel for an American
      Supersonic Transport," Paper No. 74-163, American Institute of Astronautics and
      Aeronautics.

13.   S. Weiss, "Hydrogen Fueled Aircraft," Cryogenic and Industrial Gases,
      November 1974.

14.   G.D. Brewer et al., "Study of Fuel Systems for LH_2-Fueled Subsonic Transporta-
      tion Aircraft," Report No. NASA CR-145369, U.S. National Aeronautics and Space
      Administration.

General Review Sources

Casper, M.S., editor, Hydrogen Manufacture by Electrolysis, Thermal Decomposition,
and Unusual Techniques, Noyes Data Corporation, Park Ridge, New Jersey, 1978.

Cox, K.E., and K.D. Williamson, Jr., Hydrogen: Its Technology and Implications,
Volumes 1-5, CRC Press, Inc., Boco Raton, Florida, 1979.

Dickson, E.M., J.W. Ryan, and M.H. Smulyan, The Hydrogen Energy Economy, Praeger
Publishers, New York and London, 1977.

Dyer, C.R., M.Z. Sincoff, and P.D. Cribbins, "The Energy Dilemma and Its Impact on
Air Transportation," NGT 47-003-028, Old Dominion University Research Foundation for
NASA-Langley Research Center, 1973.

Gregory, D.P., "A Hydrogen-Energy System," No. L21173, Institute of Gas Technology
for the American Gas Association, 1972.

Gregory, D.P., "The Hydrogen Economy," Scientific American, Vol. 228, No. 1, 1973.

Kelley, J.H., and E.A. Laumann, "Hydrogen Tomorrow," JPL No. 5040-1, Jet Propulsion Laboratory for National Aeronautics and Space Administration, 1975.

Mathis, D.A., Hydrogen Technology for Energy, Noyes Data Corporation, Park Ridge, New Jersey, 1976.

Veziroglu, T.N., "Hydrogen Energy", 2 Volumes, New York, 1975, Plenum Press.

Williams, L.O., "Clean Energy Via Cryogenic Technology," Advances in Cryogenic Engineering, Vol. 18, 1973.

Chapter 7

SAFETY AND ENVIRONMENTAL
CONSIDERATIONS

One of the greatest barriers to the use of hydrogen as a fuel is the common opinion that it is extremely dangerous to use and handle. This opinion is not based on facts, but on the general lack of knowledge about hydrogen. When compared to gasoline or natural gas, hydrogen presents a somewhat different set of hazards, but as a set they are not greater hazards than those of current fuels.

Gasoline, natural gas, and hydrogen can cause serious fires and explosions, but the auto ignition temperature of gasoline is 258° C, natural gas is 566° C, and hydrogen is 585° C. Thus, hydrogen is signficantly more difficult to ignite with heat than gasoline and is about the same as natural gas. Gasoline is slightly poisonous and usually contains benzene, a known carcinogen; hydrogen is nonpoisonous. On combustion, gasoline can produce highly poisonous carbon monoxide, but hydrogen combustion produces only water. In the event of a destructive accident, splashed gasoline will present a fire hazard for hours, making first aid and rescue operations dangerous and even preventing the use of rescue tools such as cutting torches to remove people from wreckage. Hydrogen, on the other hand, would volatilize and disperse within moments after the accident.

On the other hand, hydrogen is explosive with air over a wider mixture ratio and is more easily ignited by electric sparks than is gasoline or natural gas. It is completely colorless, odorless, and tasteless, making leakage detection more difficult. Liquid hydrogen is also a hazard due to its extremely cold (-252° C) temperature; on contact with air, it will freeze the oxygen and nitrogen solid, producing a momentarily explosive mixture. This frozen mixture will rapidly evaporate and dissipate under ambient conditions, but while it exists it presents a real explosion hazard.

The extremely cold temperature of hydrogen also makes it dangerous should it directly contact the skin during an accident. The extreme cold can produce severe frostbite, with resultant tissue damage, and large frostbitten areas on the human body could, of course, be fatal. Fortunately, liquid hydrogen is so cold when compared to body temperature that on contact with the skin it acts like water dropped on a red hot metal plate: it sputters and immediately bounces away (Leidenfrost effect). Because of this effect, the contact time is seldom long enough to cause tissue damage. If, however, large quantities of liquid were trapped inside a person's clothing where it could not escape, injury from cold could occur. With hydrogen, then, the greatest danger is from freezing, with a lesser danger from fire, and both of these situations are dissipated rapidly. In vehicular use, hydrogen presents a slightly different, but no greater, danger than gasoline. In home use, hydrogen will be much like natural gas.

The total lack of smell and taste of hydrogen would present some danger in the home. This could be alleviated by adding to the home-use hydrogen an odorous substance such as done with natural gas. Such chemicals can be detected by smell in incredibly low concentrations. This compound would, of course, as with natural gas, be burned to form odorless products in the hydrogen flame.

A significant number of people die each year as the result of breathing carbon monoxide produced by the partial combustion of fossil fuels. All of these deaths, whether they are caused by poorly adjusted heating furnaces or by automobile exhaust gases, are a direct result of the use of carbon-containing fossil fuels as a source of energy. Conversion to hydrogen fuel would prevent these needless deaths.

In any mobile system using hydrogen, the liquid will be stored under the positive pressure. This would allow a liquid dump to be built into the system that could be actuated by a three-axis accelerometer set to detect the violent accelerations caused by an accident. Such a system would raise the cost of fuel systems and might not be necessary, but its use would probably result in a hydrogen-powered vehicle being actually safer from fuel-related dangers than the current gasoline automobiles. Electrochemical and mass spectrographic detectors, although probably too expensive for home use, are currently available for industrial use. A hydrogen detector developed by the National Aeronautics and Space Administration based on the use of ultrasonic sound may become inexpensive enough for both home and industrial use.[1]

The U.S. National Bureau of Standards has performed several studies of the safety factors influencing the use of hydrogen as a general-purpose fuel and as the fuel for private automobiles. These studies deal with a very large number of factors, but the conclusion is that the use of hydrogen will simply require learning a few new habits and that hydrogen will not be any more hazardous than the fuels currently used.

Hydrogen use in aircraft will be much the same as its use in automobiles. In most aircraft accidents, the primary hazard to the passengers is from high-velocity impacts, with fire being of lesser importance. If a crash occurs that is minor enough that the passengers escape injury from impact, then freezing from the cold of the liquid hydrogen and fire become the major problems. Spilled hydrogen flows, spreads, and evaporates with very great rapidity, so the duration of the passengers' exposure to the hazard is not long. Hydrogen vapor at the boiling point of hydrogen, 21°K, has almost the same density as air, so that initially the vapor has little tendency to rise as it warms from this low temperature; however, it rapidly becomes less dense than air and begins to rise because of buoyancy. This tendency causes spilled liquid hdyrogen to dissipate very rapidly, because the ambient temperature is always at least 220°C above the boiling point of liquid hydrogen.

Breathing hydrogen gas has no toxic effect whatsoever. If, however, the concentration is so high that the oxygen partial pressure is reduced, hydrogen can cause suffocation. The first noticeable distress from low oxygen partial pressure depends on the age of the person, the condition of the lungs, and the amount of activity being engaged in. When resting, most people are unaffected by the reduction in oxygen partial pressure to about one-half the sea-level value. This is equivalent to an altitude of between 3000 and 4000 meters. To obtain a reduction of oxygen partial pressure to one-half standard pressure, the air would be diluted with 50 percent hydrogen. This mixture, which most people would be able to breath without discomfort, would have a density of only 53-percent air, so it would be very buoyant, with a lift of over 600 grams per cubic meter. This great buoyancy would cause it to rise away from the exposed person very rapidly.

A hydrogen fire will of course be dangerous, but probably less so than a fire fueled with burning hydrocarbon fuels. The buoyancy of both the hydrogen and its combustion products will drive the fire more nearly straight up than is the case with a hydrocarbon fire. There is very little radiation from a hydrogen fire, and damage and secondary fire from radiant heat transfer is almost nonexistent. Hydrogen fires do not produce blinding smoke or toxic combustion products such as carbon monoxide. The lack of heat, smoke, and toxic fumes would make rescue operations in the presence of a hydrogen fire less hazardous to both the rescuers and the people being helped. The question of whether a hydrogen fire is more likely or not to occur is very difficult to answer. The higher thermal ignition temperature of hydrogen would appear to provide some safety margin, but even though hydrogen is more difficult to ignite thermally than hydrocarbon fuels, it is much more easily ignited by electrical sparks. It is not clear which effect is the most important in the case of an aircraft crash.

As is the case with automobiles, hydrogen use in aircraft does not appear to be any greater a hazard than the currently used fuels. It may even be slightly less, and it will present a somewhat different hazard profile than the current common fuels. Practice, training, and good design should provide a hydrogen-fueled aircraft at least as safe as the current generation of aircraft, and probably safer.

Hydrogen has unique characteristics as a low-pollution fuel, when compared to the fuels now in use or under consideration for the future. When hydrogen is burned, the only material produced in quantity is water, effectively the same water from which the hydrogen was produced. The recycle time for this process is quite short. The water vapor that escapes from the hydrogen-fueled engine immediately enters the atmospheric pool from which precipitation is produced and thus enters the worldwide water cycle. The amount of water involved is remarkably similar to the amount obtained by the combustion of fossil fuels. One kilogram of hydrogen produces an amount of combustion energy equal to 3 kilograms of light hydrocarbon fuel. The combustion of 1 kilogram of hydrogen produces 9 kilograms of water, and the combustion of 3 kilograms of light hydrocarbons produces 4.5 kilograms of water and 11 kilograms of carbon dioxide. When used as the only fuel for a combustion engine, hydrogen produces only twice the water that would be produced to generate the same amount of energy by combustion of a light hydrocarbon fuel. In addition, the water and carbon dioxide produced by fossil fuel combustion is material that has not been in circulation in the biosphere for millions of years. The water produced by the combustion of hydrogen is a current part of the biosphere, and its use as a source for hydrogen production removes it from circulation for only a short time.

The use of hydrogen produced from non-fossil materials as a portable energy source will eliminate the fossils as sources of carbon dioxide and other air pollution. Air pollution is the result of the incomplete combustion of fossil hydrocarbon fuels and the combustion of non-hydrocarbon impurities found in these fuels. It is a greater problem than the damage to soil and water caused by extraction, because of its effect on health and vegetation and its damage to buildings and other outdoor constructions, the cost of which is reckoned in billions of dollars per year, millions of hours of ill health, and some actual loss of life. At the same time, if the combustion of fossil fuels ceased, the air would be washed clean and pure by natural processes in just a few weeks.

Most of the chemical compounds implicated in air pollution--nitrogen oxides, carbon monoxide, unburned hydrocarbons, sulfur dioxide, ozone, peroxyacetyl nitrate--are unstable or water soluble. The unstable components decompose to relatively harmless compounds a few days after they are formed. The stable components, such as sulfur dioxide, dissolve in rain and are washed from the air by precipitation. The actual substances responsible for air pollution are thus quite transitory, and the only reason they are a problem is that they are manufactured in huge quantities on a daily basis. If the use of fossil fuels ceases, there will be no more air pollution. Table 7-1 shows possible fuels and the air pollution produced by use of these fuels.

One form of air pollution, carbon dioxide, is largely ignored by most authors because it is an essential constituent of the atmosphere and currently present at a concentration of only 325 parts per million. Current theories indicate that, when the planet was young, the atmosphere contained a high percentage of carbon dioxide. The evolution of photosynthetic plants, with their ability to use solar energy to form hydrocarbon compounds and release oxygen, started a billion-year-long cycle of carbon removal from the atmosphere. This removal cycle is responsible for both the fossil fuel deposits, the oxygen in the air, and the current low level of carbon dioxide in the atmosphere.

In its current state, the natural carbon cycle of the earth operates as follows: Carbon dioxide is produced by the decay of vegetation; respiration of animals; volcanoes; and the natural weathering of coal, oil, and limestone.[2] It is removed from the atmosphere by the growth of plants, which die and are formed into peat, lignite, coal, and petroleum, and by the deposition of new limestone in the oceans. Estimates indicate that the carbon dioxide concentration of the atmosphere has remained nearly constant over the last several million years, at a value of about 290 parts per million. The carbon cycle has been in equilibrium, and the removal of carbon dioxide exactly balances the production. The coal and oil deposits currently mined for fuels are part of this large-scale billion-year-long circulation in the carbon cycle.

In the late 1800s, civilizations started using fossil fuels for the production of energy. This reintroduced carbon into the atmosphere that would not, in the natural course of events, have been released for many millions of years. The result is that the premature release of this carbon has overwhelmed the natural systems for removing carbon from the atmosphere. As a result, the concentration of carbon dioxide has increased about 35 parts per million in the last 100 years, a 12-percent increase. This may not sound like a large amount, but 189 billion metric tons sounds incredible. To attempt to get a perspective on this amount of carbon dioxide, a number of things

Table 7-1. Fuels and their Combustion Products

Fuel	Composition	Toxicity of Products* Fuel-rich Combustion	Toxicity of Products* Fuel-Lean Combustion
Manufactured			
Hydrogen	H_2^0	H_2O^0	$H_2O^0 + NO^3$
Ammonia	NH_3^3	H_2O^0	$H_2O^0 + NO^3$
Hydrazine	$N_2H_4^5$	$H_2O^0 + NH_3^3$	$H_2O^0 + NO^3$
Fossil			
Methane	CH_4	$H_2O^0 + CO_2^1 + CO^3 + A_1^4$	$CO_2^1 + NO^3$
Natural gas	$CH_2 + C_2H_6 + C_3H_8 + CO_2 + N_2$	Above + A_3^3 A_2^6	Above
Bottled gas	$C_3H_8 + C_4H_{10}$	Above	Above
Gasoline	$CH_{2.2}(C_{5-10}) + Aromatics^1$	Above + A_4^3	Above
Gasoline/lead	$CH_{2.2}(C_{5-10}) + Pb^6 + Br^5$	Above + $Pb^6 + Br^5$	Above + $Pb^6 + Br^5$
Fuel oil	$CH_2(C_{11} - C_{25})$	Above	Above
Bunker oil	$CH_{1.5}(C_{15-30}+ Aromatics)$	Above + SO_2^4 $Se0_2^6$ SH_2^3 SeH_2^6	Above + SO_2^4 $Se0_2^6$ SO_3^5
Coal	$CH_{.1} + Sulfur + Selenium^6$ $+ Ash$	Above + $Si0_2^3$	Above + $Si0_2^3$
Soft coal	$CH_{.2} + Sulfur + Selenium^6$ $+ Ash$	Above	Above

Toxicity (Amount of Gas in Atmosphere Tolerable - Superscripts)

0. Nontoxic	4. 1 to 10 ppm	A_1 Aldelhydes
1. 1000 to 10,000 ppm	5. 0.1 to 1.0 ppm	A_2 Unsaturated Hydrocarbons
2. 100 to 1,000 ppm	6. 0.01 to 0.1 ppm	A_3 Unsaturated Aldelhydes
3. 10 to 100 ppm		A_4 Aromatic Hydrocarbons
		* Plus unburned fuel

must be considered.   An increase of 35 parts per million has no direct effect on people, even those in poor health; a total concentration of 3000 to 5000 parts per million would be required.   If, however, the concentration of carbon dioxide in a person's blood increased by 12 percent, the individual would be in clear disteress; so, at least for some living systems, a 12-percent increase in carbon dioxide is detrimental.

Tests have also shown that an amount of carbon dioxide about equal to that found in the air has been added to the first 30 or so meters of the ocean.   (Virtually all the carbon dioxide that man has added to the atmosphere could potentially be dissolved in the ocean, if the entire ocean body was available.   Unfortunately, the mixing time for the oceans is many thousands of years, so that in the 100 or so years that man has been adding carbon dioxide, only the top 30 meters have been involved).   The carbon dioxide dissolved in water makes it acid.   Predictions indicate that if fossil fuel combustion is continued at the current projected rate, the ocean's top 30 meters will become so acid that shellfish will no longer be able to produce shells and will die.   It is predicted that, with the current rate of carbon dioxide addition, this will occur in 2010.[3]   Along with the specific loss of the shellfish, there will be profound effects on the total ocean food chain, cascading through the whole life structure of the ocean and profoundly affecting all forms of life.

Carbon dioxide strongly absorbs infrared radiation, which is the heat portion of the spectrum.   The earth radiates heat to space at an infrared wavelength of 1 micrometer.   Making the atmosphere less transparent at these wavelength causes more heat to be retained on the earth's surface and this should increase the temperature of the earth.[4]   Particulates in the atmosphere, often placed there by the same combustion process that introduces carbon dioxide, block sunlight, making the earth cooler.   Over the last 20 to 30 years, these two effects seem to have nearly cancelled out in the Northern Hemisphere, and temperatures have not been observed to be increasing.   The process that mixes the air between the two hemispheres apparently removes only the particulates, allowing the carbon dioxide warming to occur in the Southern Hemisphere, where the average temperature has been observed to increase a few degrees over the last 20 years.

These temperature effects are serious in their own right, but such changes in the heat balance of the earth can also cause shifts in weather patterns.   There have already been clear indications that weather patterns are changing.   North Africa, England, India, and the Western United States have been having unusually dry weather, and data indicate that the average direction and velocity of the wind has been changing in England.   Thus far, these effects are relatively small, but they have caused severe hardship for the populations affected.   If they are a result of disturbing the heat balance of the earth by the increase in carbon dioxide in the atmosphere from the combustion of fossil fuels, it is imperative that carbon dioxide production is reduced or eliminated as soon as possible.[5]

There has been speculation that plants will remove some of this carbon dioxide from the atmosphere and will prevent the total increase from becoming great enough to cause problems.   Several factors, however, prevent this.   In controlled experiments, plants respond to increased carbon dioxide by increased growth rates, but in most natural situations, the growth of plants is limited by factors such as water, nitrogen, and phosphorous supplies, such that plants are unable to respond to the increased availability of carbon dioxide[2].   In addition, the carbon dioxide removed by plants is only temporarily removed from the atmosphere.   Most of it is returned to the atmosphere by decay processes in a few years.   Only in places where coal or pre-coal peat deposits are being formed is the carbon dioxide being removed from the atmosphere in a permanent sense.   The process of coal formation is quite slow compared with the rate at which fossil fuel carbon dioxide is being added to the atmosphere, and we cannot count on its help in reducing the buildup.

The oxygen by-product resulting from the production of hydrogen by electrolysis or thermochemical processes can be used in pollution abatement.   This topic is not directly related to the use of hydrogen as a fuel, but the indirect effect of the availability of large quantities of oxygen may have a significant impact on the overall desirability of the hydrogen power system.   Because of its potential contribution to the acceptance of hydrogen power, a brief discussion seems appropriate.

In the treatment of sewage, several processes are used that depend on the ability of bacterial respiration of oxygen for the oxidation and reduction of the waste materials in the water.   In water treatment plants operating with this process, the sewage is placed in large concrete holding tanks and air is bubbled into it.   The bacteria use the oxygen from the air for respiration, growth, and multiplication as they consume the organic substances in the water.   The products of this process are primarily water and carbon dioxide formed by the bacterially catalyzed oxidation of the materials in the water.   The parameter that limits the rate at which the bacteria can oxidize the organic waste is the rate at which the oxygen from the air can be transferred from the gas phase bubbles into solution in the water, where it is accessible to the bacteria.   The rate at which the oxygen from the air can enter the water is in turn limited to the partial ·pressure of the oxygen in the air.   Air contains 21 percent oxygen so that the maximum partial pressure of oxygen is 0.21 atmosphere in an open tank. Two methods are available to increase the partial pressure of the oxygen: One is to increase the total pressure in the tank, the other is to use pure oxygen. Increasing the total pressure would require that the tank be sealed and built strong enough to resist the desired pressure, a very costly type of tank to build compared to a simple open concrete tank built much like a swimming pool. If pure oxygen is used in this same open tank, the partial pressure of the oxygen is one atmosphere, or about five times higher than the partial pressure obtained by using air. In practice, the rate at which a given size tank can process sewage, when pure oxygen is used instead of air, is about five times higher. This gain in efficiency can be used to build less costly sewage disposal plants or plants that do a more thorough cleaning of the input water.[6]

With the very large quantities of low-cost oxygen available on implementation of hydrogen power, it would be economically feasible to attempt to reverse some of the damage already done to the environment. For example, large quantities of oxygen could be fed directly into rivers such as the Hudson and Detroit. This would strongly increase the rate of aerobic bacterial purification of these waters. In the case of the Detroit River, large-scale and massive oxygenation would ultimately result in improvement of the quality of water in Lake Erie. In the same manner, oxygenation of the effluent waters that ultimately end in the Mississippi River, the Rhine, the Seine, and others, and direct oxygenation of the rivers and their tributaries, would greatly improve the quality of the water all the way to the sea.

Several pilot operations using this principle are in operation. The Linde Division of the Union Carbide Corporation has developed a system in which water is pumped from a river or lake and caused to flow through a pipe under high pressure. Pure oxygen is injected into the high-pressure pipe, where it dissolves in the water. This process allows a much greater volume of oxygen to be dissolved because of the high pressure. Finally, the water, which is supersaturated with oxygen at ordinary pressures, is injected into the body of water from which it was removed. This oxygenated water mixes with the other water, greatly increasing the rate of degradation of any organic chemical pollutants. In addition, the oxygen is available to support higher forms of aquatic life.

In several of these pilot operations, a small percentage of the oxygen has been converted to ozone before it is injected into the water. Ozone is an extremely active form of oxygen. Because of its extreme reactivity, it is toxic to all life forms, but this reactivity allows it to act on many of the pollutants that are only slowly acted on by the natural oxidation process. When ozone acts on such difficult-to-degrade materials as biological hard detergents, it produces chemicals that can be further degraded easily by the natural cycle. This reaction destroys the ozone and, with it, its toxicity.[6]

Ozone must be used with great care to ensure that the toxicity introduced by its presence is not more of a problem than the material it was introduced to eliminate. Its use would therefore be limited to the elimination of specifically difficult-to-degrade materials in well-studied and monitored situations. Even without the presence of organic pollutants, ozone decomposes to oxygen at a fairly rapid rate, so its use would not result in the buildup of a new toxin in the environment.

In existing pilot operations using oxygen enhancement, for example, the Crown Zellerbach plants in Calmas, Washington, and Bogalusa, Louisiana, for paper waste treatment and the Batavia, New York, sewage disposal plant, difficult-to-degrade organic materials have been quite significantly reduced, and increases in dissolved oxygen, so important to aquatic animals, have been observed many miles downstream from the treatment point. Using conventional sources of oxygen, Union Carbide has projected a savings of 30 to 50 percent in the treatment of sewage water.

In incinerating solid waste, the most severe problems are associated with incomplete combustion of the waste. Oxygen can also be used to eliminate this problem. Cities can construct or modify incinerators using an oxygen-fed fire, with two advantages. First, the flame temperature and chemical reaction rate are much higher, allowing a much higher rate of waste processing in a given size incinerator. Second, the oxygen can be kept in excess, preventing the formation of incomplete combustion products. Such oxygen-fired incinerators would not produce the carbon monoxide or organic chemical smog of incomplete combustion produced by conventional burning. These incinerators would produce a mineral fly ash that could be collected by conventional techniques of cyclone separation and electrostatic precipitators. They would also produce sulfur dioxide and chlorine if these were present in the waste, as they usually would be. These materials can be removed by chemical means such as absorption in lime and can often be converted to by-products for sale to the basic chemical industry. They can be recovered from the exhaust gas from the oxygen-fired incinerators much more economically than from conventional air-fired incinerators because the exhaust gas would not be diluted by a factor of 4 to 1 with atmospheric nitrogen.

Oxygen-enhanced combustion would be advantageous in the operation of power plants used to produce power by the combustion of trash. The oxygen combustion would produce higher temperatures, increasing the generating efficiency. The high intensity and temperature of this type of combustion process would also make it possible to process waste trash and heavy scrap such as discarded machinery and automobiles in the same processing unit. The high temperature and oxygen will burn such organic materials as rubber, paint, and upholstery, and the heat from this material plus the heat from the burning refuse will melt the metal. This melted metal can be cast into ingots for use in noncritical metal applications or processed as scrap metal to produce high-quality metal. For a large city, the average composition of daily refuse is relatively constant, so the metal produced by processing the refuse would be uniform in composition. The composition would, however, be rather unusual, for it would contain mostly iron, with its usual alloying ingredients, plus copper, zinc, tin, and aluminum from electric wiring, die-cast parts, tin cans, and aluminum from both autos and discarded household utensils. This alloy could probably be used as is for many noncritical applications, or it could be purified by conventional techniques to produce various high-quality iron and steel alloys, with the other metals separated for recycling.

A technique demonstrated by the Pittsburgh Coal Research Center, Bureau of Mines, consisting of the pyrolysis of trash with excess hydrogen, should be useful for solid waste processing. In this technique, trash and hydrogen are placed in a closed container and heated to a temperature of 300° to 600°C. The organic carbon compounds react to form methane and other natural gas and petroleum type compounds that could be used as raw materials in plastics manufacturing and similar processes.[7] The glass would be recovered as slag, and most of the metals would be recovered as free metal. Development of this technique, in conjunction with the large-scale use of hydrogen, would allow the reduction and elimination of solid waste to produce an economically valuable by-product.

Chapter 7 References

1.   "Acoustic Hydrogen Detector," Industrial Research, September 1968.

2.   B. Bolin, "The Carbon Cycle," Scientific American Vol. 223, No.3

3.   A.W. Fairhall, "Effects of Carbon Dioxide on Ocean Life Systems," Nature, Vol. 245, September 7, 1973, p. 20.

4.   A.H. Oort, "The Energy Cycle of the Earth," Scientific American, Vol. 223, No. 3.

5.   M. Glantz, "A Political View of CO_2," Nature, Vol. 280, July 19, 1979.

6.   "O$_2$ and O$_3$-Rx for Pollution," Chemical Engineering, February 27, 1970.

7.   R.C. Corey, "Pyrolysis, Hydrogenation and Incineration of Municipal Refuse," Proceedings of the Second Mineral Waste Utilization Symposium, Illinois Institute of Technology, Chicago Illinois, March 1970.

General Review Sources

American Society of Safety Engineers Journal, "Liquefied Hydrogen Safety - Review," Vol. 14, No. 5, May 1969.

Arvidson, J.M., J. Hord, and D.M. Mann, "Efflux of Gaseous Hydrogen or Methane Fuels from the Interior of an Automobile," U.S. National Bureau of Standards Technical Note 666, 1975.

Edeskuty, F.J., and R. Reider, "Liquefied Hydrogen Safety," PB-230845, Technology Applications Center, University of New Mexico, Los Alamos Scientific Laboratory, New Mexico, 1968.

Hord, J., "Is Hydrogen Safe?," U.S. National Bureau of Standards, Technical Note 690, 1976.

International Journal of Hydrogen Energy, Official Journal of the International Association for Hydrogen Energy, Published by Pergamon Press.

National Aeronautics and Space Administration, "Hydrogen Safety Manual," NASA-TM-X-52454, Lewis Research Center.

Porter, J.B., "Analysis of Hydrogen Explosion Hazards," Du Ponte de Nemours and Company, Aiken, South Carolina, July 1972.

BIBLIOGRAPHIES

U.S. Bureau of Standards, "Hydrogen - Future Fule - A Bibliography (With Emphasis on Cryogenic Technology, " Technical Note 664, 1975.

U.S. Energy Research and Development Administration, "Hydrogen Fuels - A Bibliography, TID 3358, 1976.

Hydrogen Energy Bibliography, published regularly in the bimonthly journal of International Journal of Hydrogen Energy, Pergamon Press.

INDEX

157